SURFACE ANATOMY

SURFACE ANATOMY

JOSEPH ROYCE, Ph.D.

Associate Professor
Department of Physical Education
University of California, Berkeley

F. A. DAVIS COMPANY, Philadelphia

Printed in the United States of America

Library of Congress Catalog Card Number 65-22235

Preface

This volume is not meant to be a substitute for a complete course in anatomy, physiology, kinesiology, human growth or morphology. It is, rather, a book for students of the human exterior. It presupposes a ready knowledge of the body interior, but attempts to assist those who, because of their work or interest, like to complement their knowledge so that they can identify quickly the visible signs of muscle function. Too often the complaint is heard that a thorough study of anatomy, using cadavers, tends to remove the anatomist from reality. Confronted with a living model the question "what bulges there?" must often remain unanswered. The artist, on the other hand, has worked with so many nude models that he knows how to draw them, although he might be unable to understand why certain muscles bulge during certain actions or how to make them contract at will. The kinesiologist or physical therapist, again, would like to know what muscles are involved in the action of which he has made a movie and thus requests guidance in the interpretation of surface contours.

This book represents a modest attempt to lend a helping hand to students of anatomy, physiology, kinesiology, physical therapy and art in bridging the gap in awareness that seems to exist whenever human models are used sparingly during the years of instruction. Ideally, the observer develops a keen eye for the interesting differences and commonalities in human forms and actions. A thorough knowledge of the human exterior elevates the observation of human beings at work, play or rest to the level of an art. The anatomist, kinesiologist or physical therapist, thus familiarized with the wide variations in composition, contours and muscle function of man in action, may become a knowledgeable observer who will instantaneously become aware of interesting features, assets or deviations in "what walks by." This high awareness can only lead to increased knowledge, interest and pleasure.

Some books dealing with surface anatomy leave the reader with the impression that most people reveal clear-cut skeletal landmarks or muscle contours. This impression, however, is erroneous. On the contrary, owing to normal fat padding, it is often extremely difficult to find a specific surface muscle or a bony landmark. That most surface muscles are photographically represented in this volume is due to the large number of subjects (each having a specific point of interest) used to obtain some 2000 photographs from which the ones appearing in this book were chosen. No professional model is included; the subjects used were friends and neighbors—to leave the impression that the photographs are those of *real people,* like the boy or girl next door.

Although the photographs customarily found in textbooks are usually acceptable from a technical angle, they are often unsatisfactory from an aesthetic viewpoint. There is no reason why a good, informative photograph cannot also be beautiful. Instead of showing the naked chest of a woman who stands at attention with sagging stockings and rolled-down skirt, it might be more aesthetic to present a completely nude figure in a relaxed standing position. If, moreover, the aloha curtains are replaced with a plain black backdrop one has gained in simplicity and pictorial beauty. Now if the irritating eyepatches and fig leaves are removed and the subject is posed properly or the print is cropped to make the subject unrecognizable, a picture that is natural and artistic *as well as* informative may be obtained.

The camera used for taking the photographs shown in this book was a 35 mm Exacta V. For most of the pictures, three floodlights were used: one on each side of the subject at a distance of six feet (providing striking light), and one in front of the subject behind the photographer (for fill-in purposes). Deviations from this basic setup were made as needed. Exposures were made at f 4.5 and $\frac{1}{100}$ second. The film employed was Kodak Plus X, which was developed in Microdol and enlarged on Kodak Polycontrast F paper.

The subject was instructed to perform a precisely outlined action and was then turned to a position which gave the desired shadow masses. To "shoot" a specific muscle becomes difficult because many muscles act in groups. First the photographer may see the desired muscle swell, but after less than a second other muscles in the group may contract to act as assistants, thus obliterating the contrast with the first muscle. Therefore, "shooting" must often be done fast and immediately after the contraction has started. By rotating the subject slightly back and forth the photographer may obtain one photograph out of many that has the qualities wished for.

The majority of the photographs were taken by the author. Exceptions are Figures 22, 23, 29, 33, 72, 79 and 109 which were taken by Mr. Alan Stumpf of the Entomology Department at the University of California, Berkeley. Permission to use Figure 127 was generously given by Doctors L. M. Bayer (Presbyterian Medical Center, San Francisco) and Nancy Bayley (Institute of Human Development, University of California, Berkeley), and the University of Chicago Press.

The author wishes to thank all who helped with technical assistance, facilities and equipment. He would also like to express his gratitude to the numerous anonymous subjects who, knowing that the photographs of only a few of them would be chosen for this book, gave generously of their time and efforts.

Joseph Royce

Contents

	Page
Introduction	1
Photographs of Regional Anatomy	14
Photographs of the Full Figure	198
Changes in Proportions and Stance (Eighteen Months to Eighteen Years)	254
Index	257

Introduction

The skeleton may be considered as a structure of levers. Figure 1 shows the lever types that are most common in the human body. Each lever system consists of a rigid bar, rotating around a fulcrum (f), a resistance (R) and a force (M). In our body we apply the force by means of a muscle contraction to the bone, which rotates around the joint. The resistance may be provided by gravity, friction, and the like. In a complete skeleton a *series* of levers, all interconnected at the joints, makes the execution of very complex and precise motions possible.

The complexity of many movements can be appreciated by viewing the following example: have a subject flex his elbow joint slowly against the resistance of a weight held in the hand (see Figure 1A). This action takes place in the elbow joint as fulcrum, with the biceps providing the force. Now, *while* this action takes place, the *upper* arm can be raised forward, the shoulder joint being used as fulcrum while the anterior deltoid provides the force. Here we have then a lever action (forearm) with a *moving* fulcrum (elbow) *on top* of another lever action (upper arm)!

This action can be made more complex, nearly *ad infinitum,* by having the subject move his body as a whole while moving all parts simultaneously. To appreciate the beauty of coordinated movement the perceptive student of human motion should watch a slow-motion loop film of such a "simple" motion as walking. In a rhythmic pattern the limbs describe their arcs, the trunk rotates and weaves, and the skin is alive with swellings, depressions, furrows and ridges.

To be able to identify the muscles that are alternately bulging and disappearing we must know something about the action, the directions of the forces and resistances, and the lever system, as well as something about the location of the muscles. In locating the muscles we are helped by visible (or palpable) landmarks on the body. The bony landmarks especially are relatively stable in their relation to the sites of origins and insertions of the muscles, and provide good reference points. The most common bony landmarks are indicated in Figure 2.

Identifying a bony landmark on a skeleton differs markedly from identifying the same landmark on a living subject, especially if the landmark in the living subject is covered by a generous amount of adipose tissue. Often it is advantageous to estimate the position of a hidden landmark from clues such as that revealed by observation of the axis of rotation of a joint; as an example, by palpation when the knee

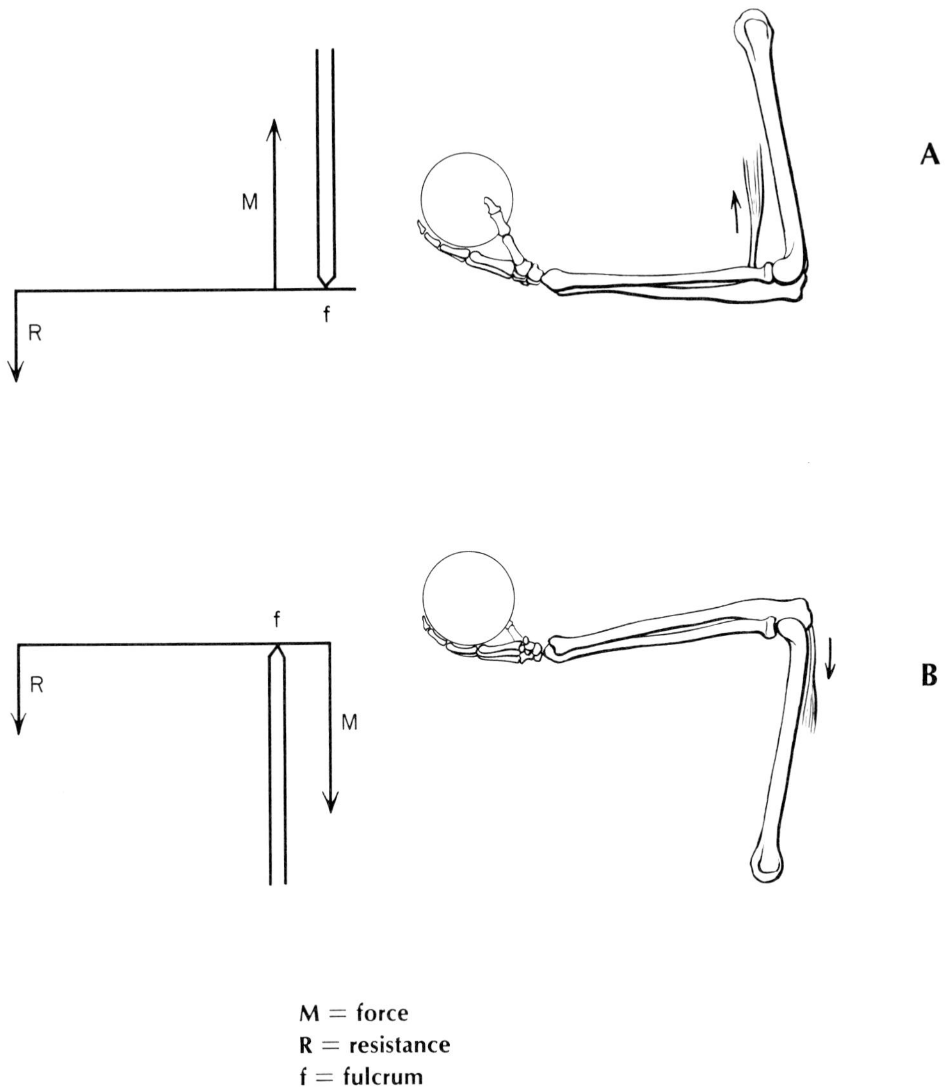

Figure 1

A. Schematic representation of the action of the biceps on the forearm resulting in lifting a weight (against gravity).

B. The triceps in a similar action.

is lifted and lowered (hip joint is flexed and extended) the greater trochanter can be found approximately on the axis of rotation in the hip joint.

After the student has become familiar with the location of the skeletal landmark he can proceed to identify the muscles that are involved in some specific action. Although it is impossible to estimate the exact contribution (in terms of force, applied for a certain time) of each muscle to a movement, it is often possible to make a muscle bulge by going through a specific motion. For instance, if the elbow is flexed against the resistance of a weight held in the hand, we see the swelling and shortening of the biceps. Pressing the hands together in front of the chest brings the pectoralis major into action. Standing with bent knees makes the quadriceps femoris contract.

The involvement in certain actions of specific muscle groups can sometimes be deduced from the knowledge of the direction of the forces. The static pose shown in Figure 2, for instance, can only be held (against the force of gravity acting downward) if the muscles in the left elbow and left knee attempt flexion while those in the right elbow and right knee attempt extension. A careful study of common static poses and a good knowledge of the action of isolated muscles might give us a clue as to which muscle groups are responsible for holding these poses and which groups should be relaxed. Inexperienced sculptors often depict figures with those muscles bulging (contracting) that could only contribute to collapse if the figure were a living model.

Careful viewing of Figures 3 through 6 may lead to some insight into the role played by the major muscle groups of the body. In Figure 3, for instance, we *know* (just look at the photograph and try this pose!) that the subject maintains this static position by contracting the extensors of the knee and hip joints and the flexors of the elbow joint. Together with the flexors of the elbow joint a great number of muscles form a chain of forces or a force-coupling which, when maintained and balanced carefully, makes it possible to *hold* the static position.

While muscle analysis of a *static* pose may become quite involved, the complexity of a *movement* requires intricate instrumentation (electromyograph, recording goniometers, accelerometers, force plates, etc.) for its analysis. Basic to such a complex analysis, however, is the skill in recognizing and locating surface muscles. For aiding the student in the acquisition of this skill the photographs in Figures 7 through 126, making up the bulk of this book, are presented. These pictures also serve a second but no less important purpose: that of helping the student to gain an insight in the composition of the body as a whole in varying individuals—in infants and adults, men and women, the rather plump and the thin. It cannot, however, be claimed that these photographs are representative of *all* types of human beings or even *specific* types. On the contrary, they depict very normal, healthy and active people, such as those who work in your office, play on the beach, walk in the shopping center, and look like *us*.

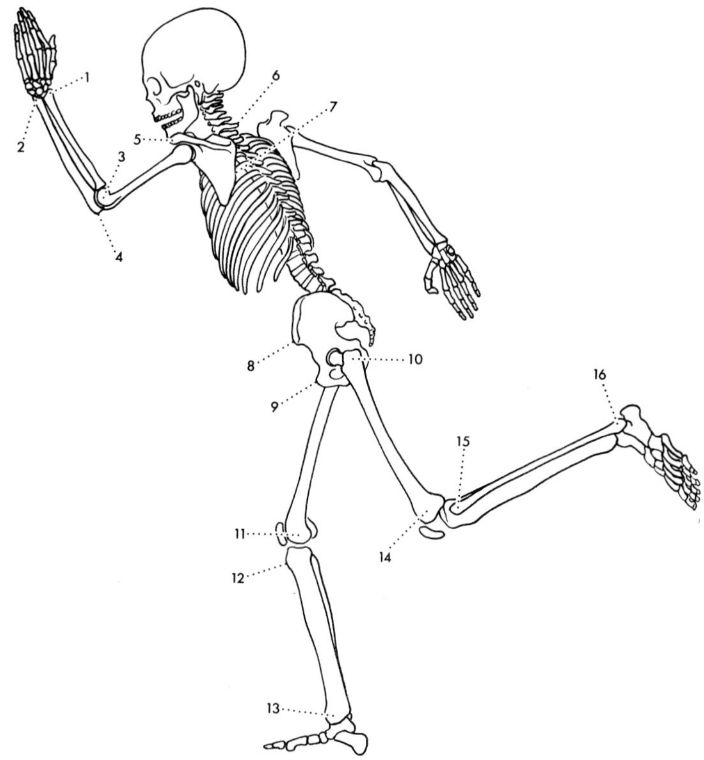

1. **Styloid process (radius)**
2. **Styloid process (ulna)**
3. **Epicondylus lateralis humeri**
4. **Olecranon (ulna)**
5. **Acromion (scapula)**
6. **Spinal process (7th cervical)**
7. **Medial border of scapula**
8. **Crista iliaca anterior superior**
9. **Pubic arch**
10. **Greater trochanter**
11. **Epicondylus medialis femoris**
12. **Tuberositas tibiae**
13. **Medial malleolus**
14. **Epicondylus lateralis femoris**
15. **Head of fibula**
16. **Lateral malleolus**

Figure 2 (♂ 11)

Skeleton on which some of the most common visible and palpable bony landmarks are indicated, accompanied by a photograph on which the approximate corresponding location is marked on a living model.

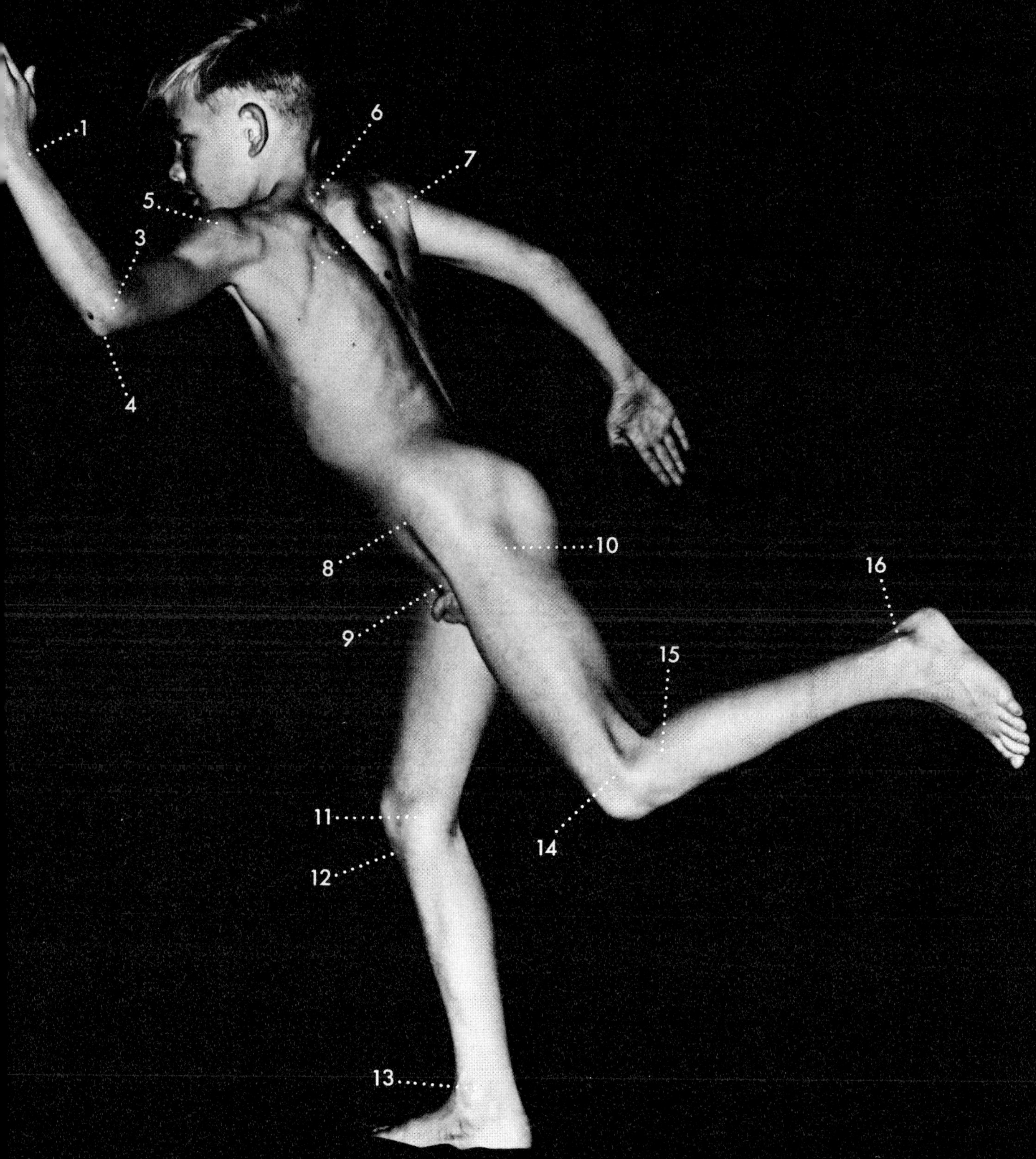
1
6
7
5
3
4
10
8
16
9
15
11
14
12
13

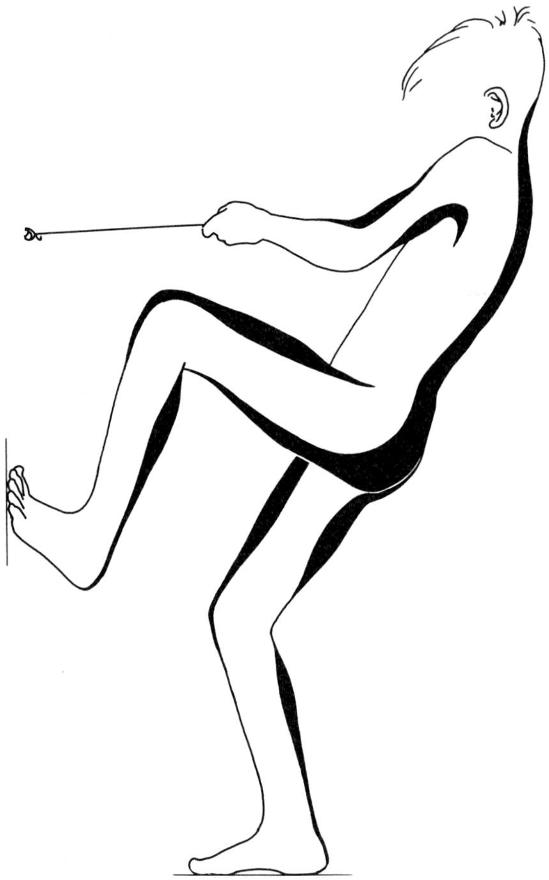

Figure 3 (♂ **11**)

Pulling on a rope. Attempts are made to flex the elbow joints, combined with an effort to extend the shoulder, hip and knee joints as well as the spine. The ankles and feet attempt to plantar flex.

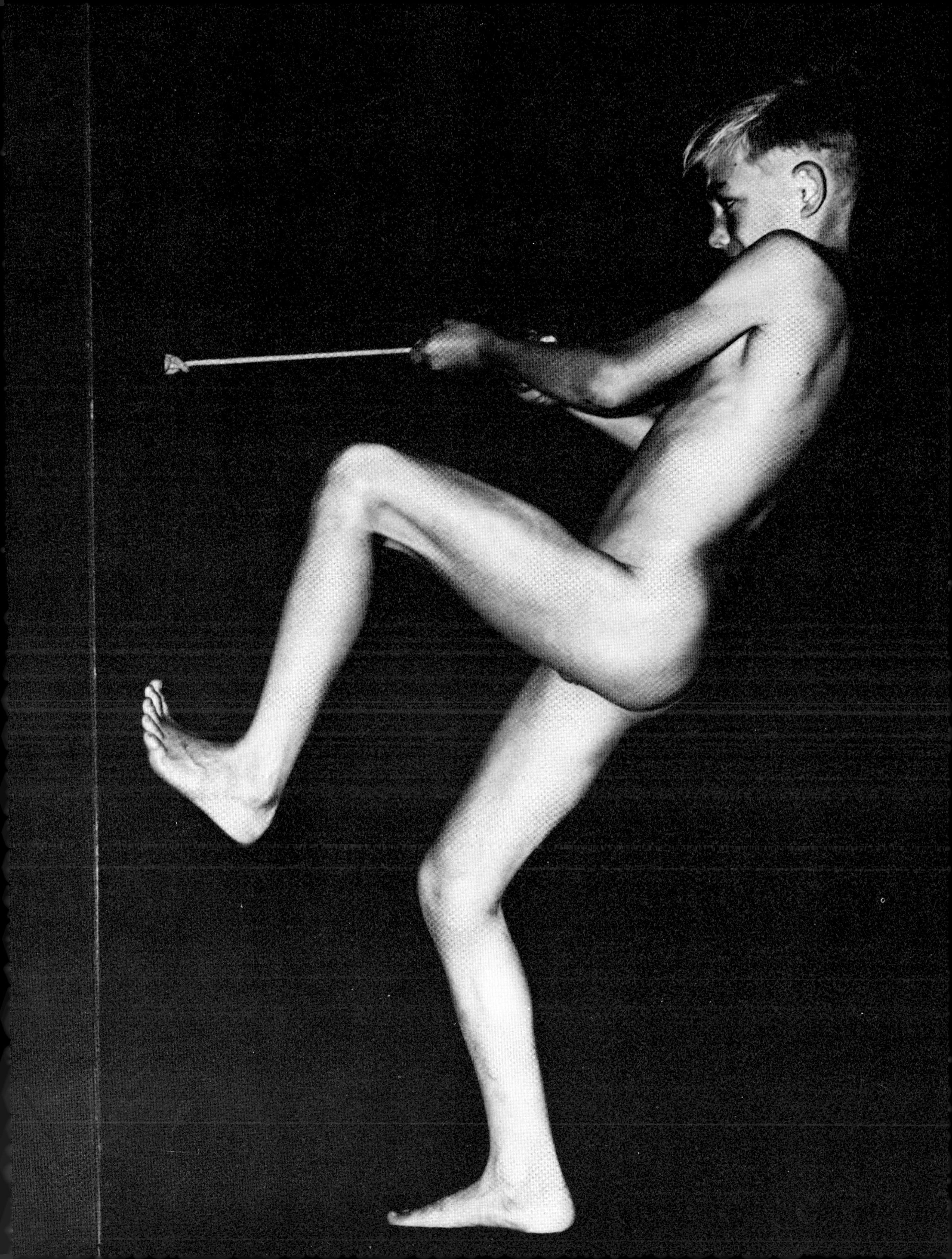

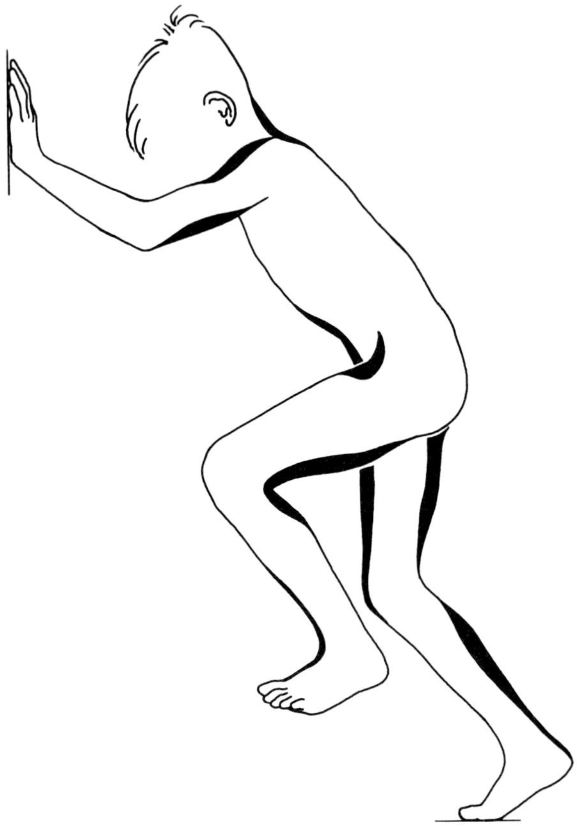

Figure 4 (♂ **11)**

Pushing a heavy object. An attempt is made to extend the neck, elbows, right hip and right knee. The shoulder joints, left hip and left knee are being flexed. The right ankle and foot attempt plantar flexion while the left ankle and foot are dorsiflexed.

Figure 5 (♂ 11)

Lifting a heavy weight. The spine, hips and knees have to be extended.

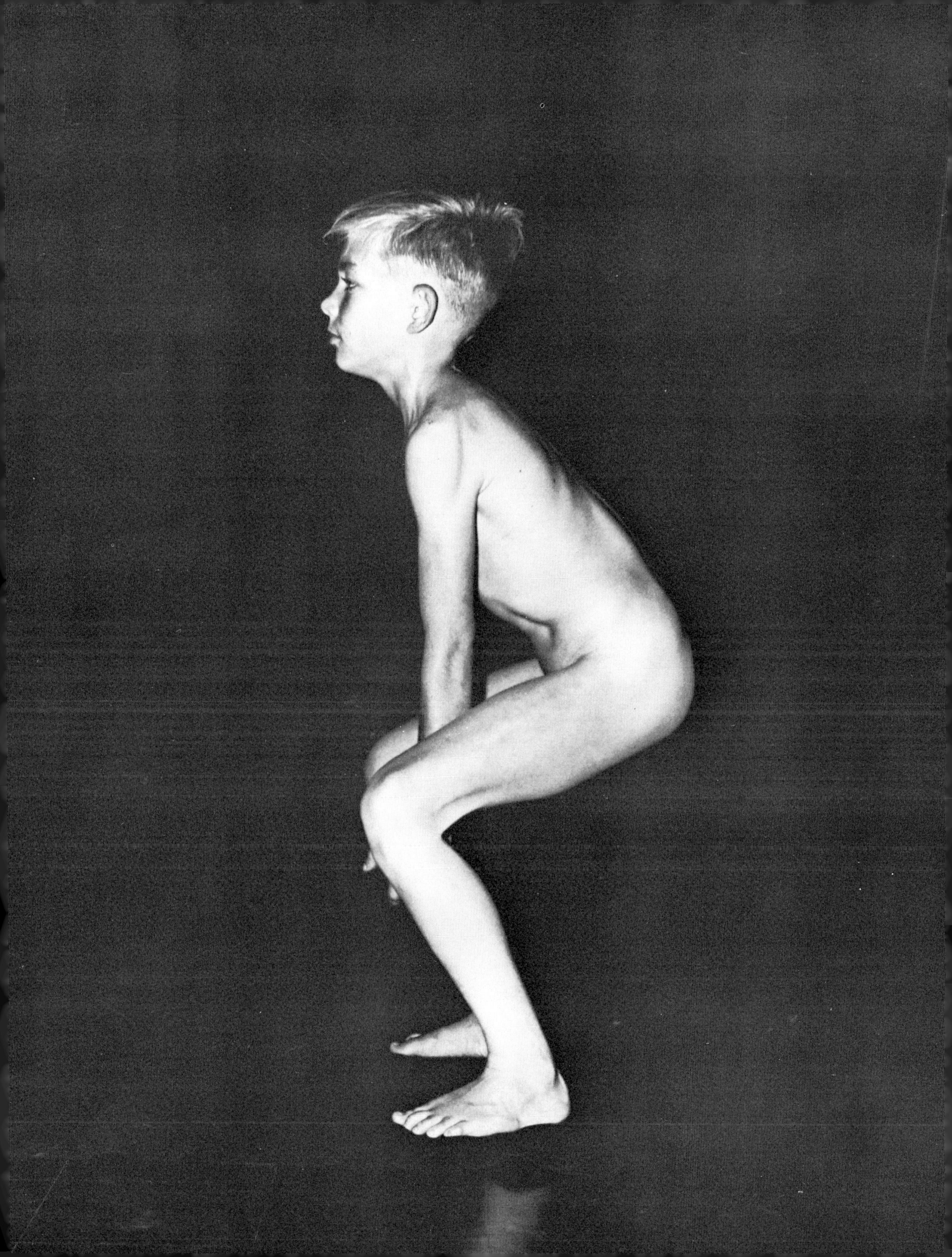

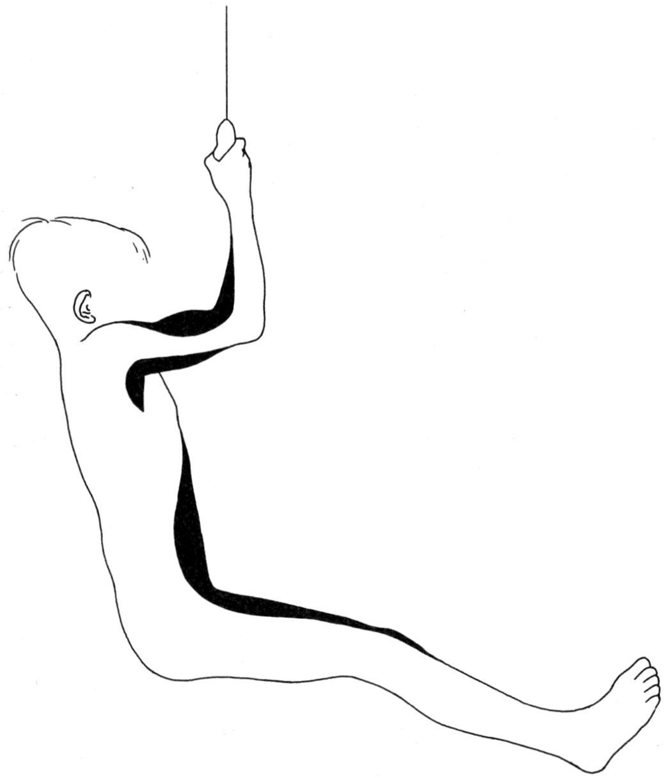

Figure 6 (♂ **11**)

Hanging on a rope. The elbows and hips are statically flexed while the shoulder joint and knees are being extended. To keep the spine straight and form a basis upon which the heavy legs may be flexed the spine is flexed.

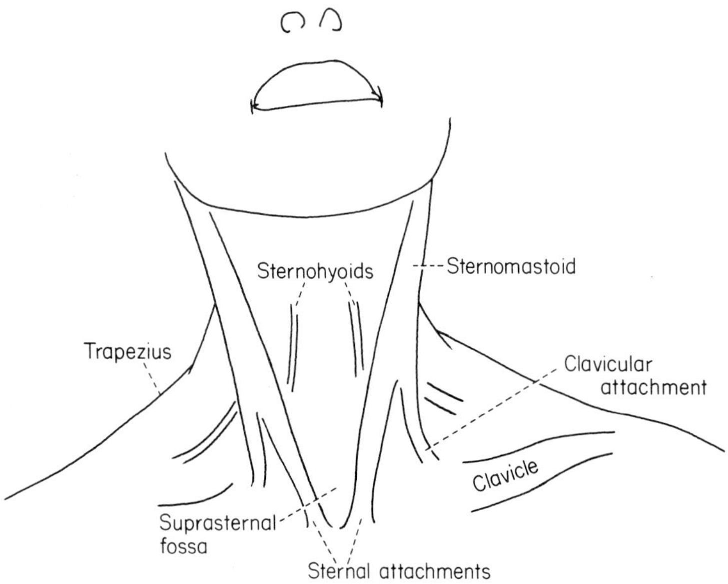

Figure 7 (♀ 26)

The most obvious skeletal landmarks shown in this figure are the zygomatic arch (cheekbones), the angle of the mandible (jaw), and the clavicle.

The head is supported by the cervical column and maintains its rather unstable equilibrium because of the action of muscles. When the head has been moved slightly backward its center of gravity moves beyond the point at which it is supported. By contracting the sternomastoid muscles on both sides of the neck the head can be kept from "rolling" further backward. Contraction of only one of these muscles turns the face upward and to the opposite side of the contraction (Fig. 8). This is easily demonstrated if the action is resisted.

The sternomastoid is a very prominent muscle and is easily seen extending from behind the ear downward and, after splitting up, inserting at the sternum and the medial part of the clavicle.

The border of the upper part of the trapezius is clearly demonstrated by an attempt to shrug while the shoulders are kept down by holding weights in the hands (see also Fig. 19). The steepness of the slope of the upper trapezius is generally more prominent in males than in females because of better muscle development in the males.

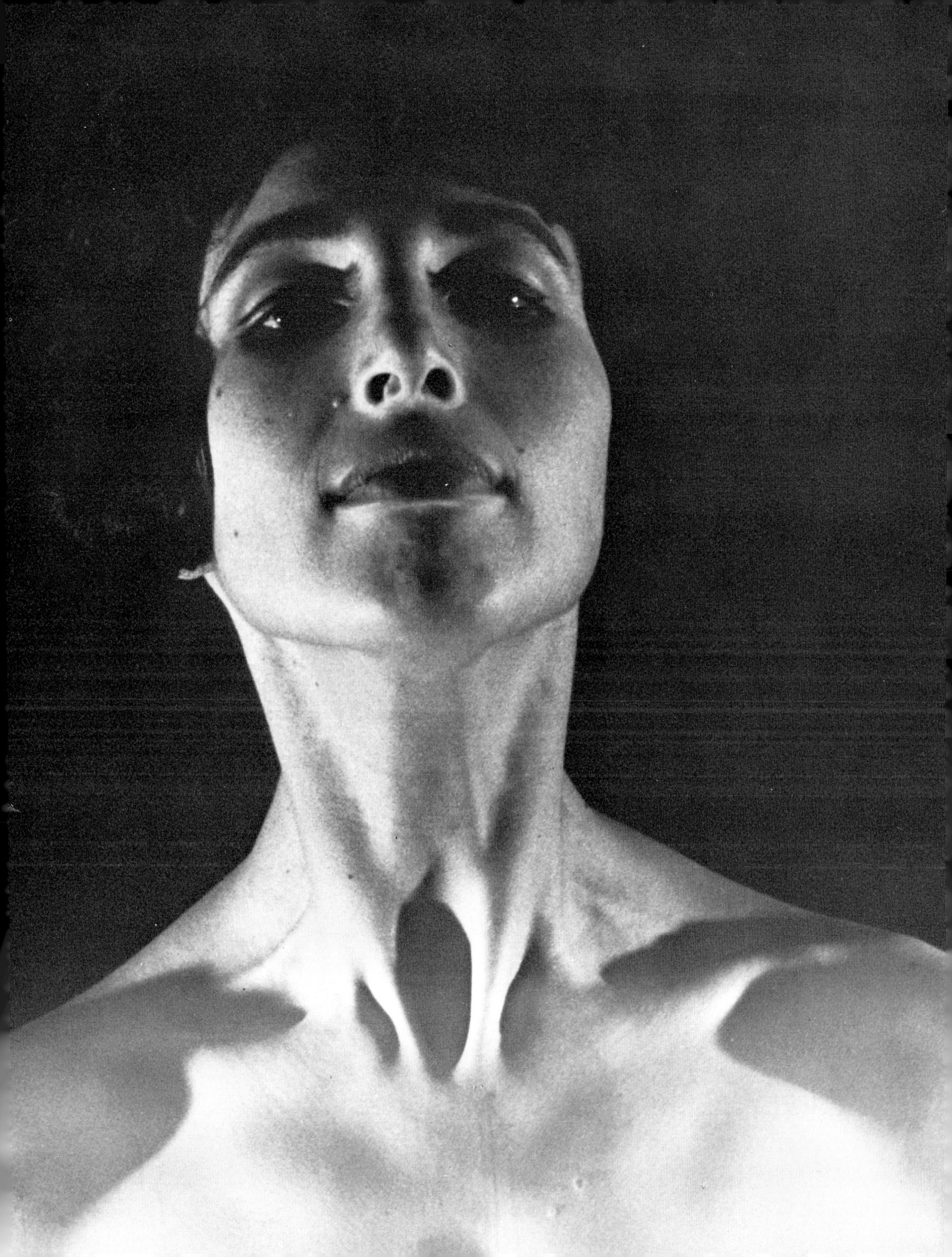

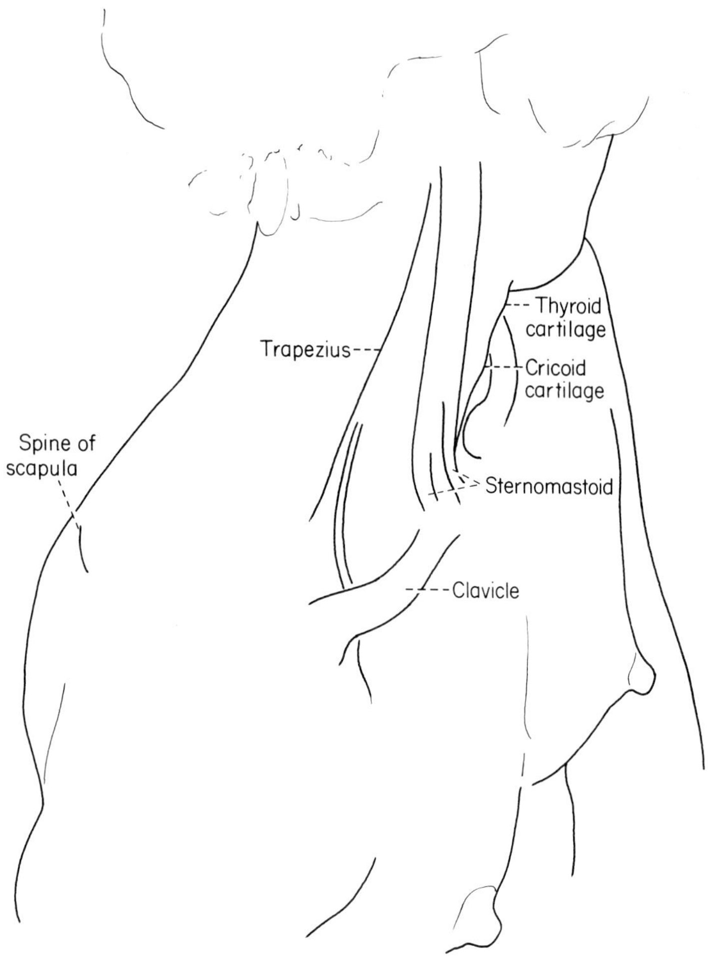

Figure 8 (♀ 26)

The sternomastoid (rotating the face away from the contracting side) and the upper trapezius together with the clavicle form a large triangular area —the greater supraclavicular fossa.

The scapula and the clavicle are joined at the acromion which can be palpated easily. The only skeletal attachment of this shoulder girdle is by the sternoclavicular joint, thus a great deal of mobility is preserved. Without this mobile basis for the humerus to act on, the range of movement of the arm would be restricted greatly.

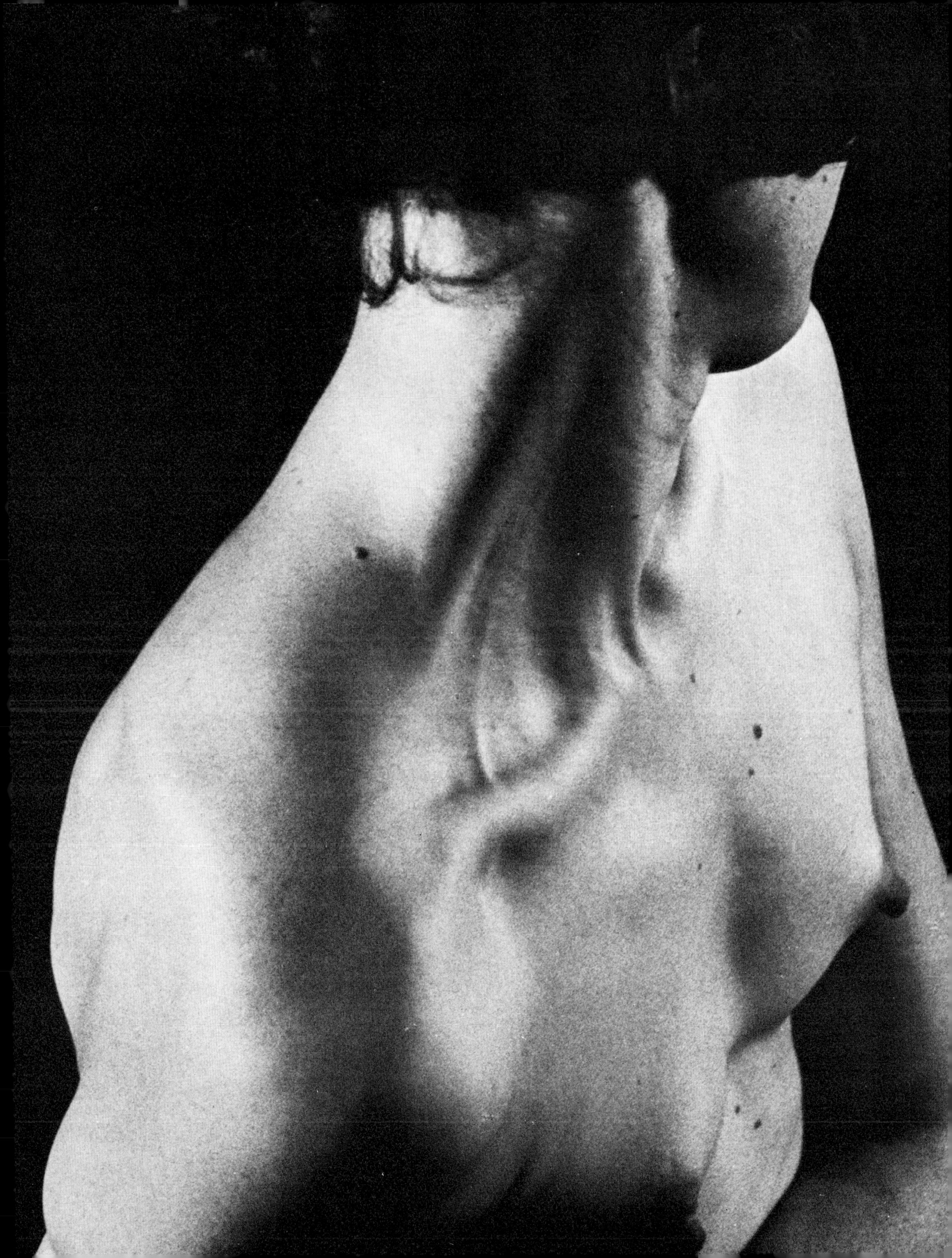

Figure 9 (♀ **11**)

A thin sheet of muscle, the platysma, originates from the fascia pectoralis on the shoulder and in the breast region. Its contraction pulls the skin of these areas upward and away from the underlying tissues in the neck. It also draws the jaw downward.

The oval-shaped area around the mouth indicates the orbicularis oris muscle, a ring muscle that closes the mouth and points the lips.

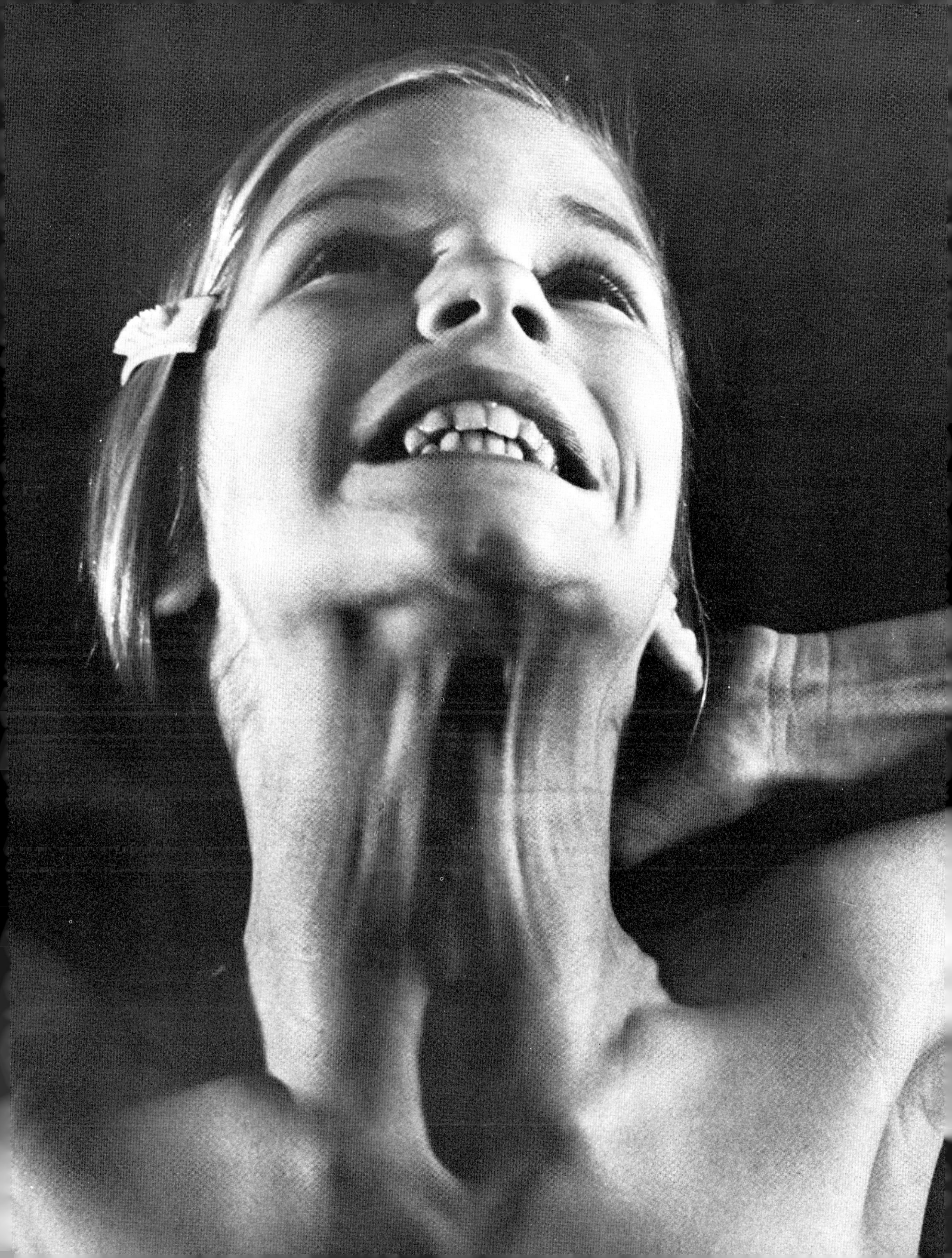

Figure 10 (♀ 11)

The lower attachment of the sternomastoid (sternal part) is centered on top of the manubrium (the "handle" of the sternum). The whole outline of the clavicle is discernible. Below the lateral third of the clavicle the coracoid processes can be seen on people with a thin overlying pectoralis major. Bringing the elbows backwards may demonstrate this bony landmark.

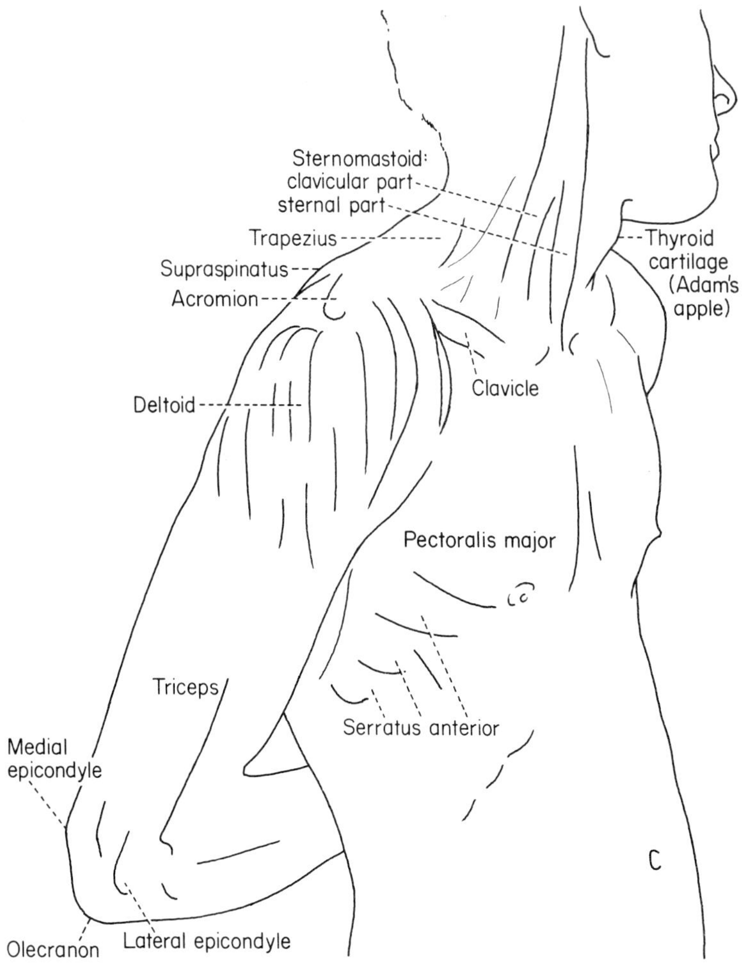

Figure 11 (♂ 23)

The contours of neck and shoulder may be demonstrated by hooking the fingers of both hands behind the back and pulling against each other. With the head turned sideways and the shoulders pushed downwards many muscles can be tensed so as to become clearly visible.

The deltoid shows its many distinctive strands.

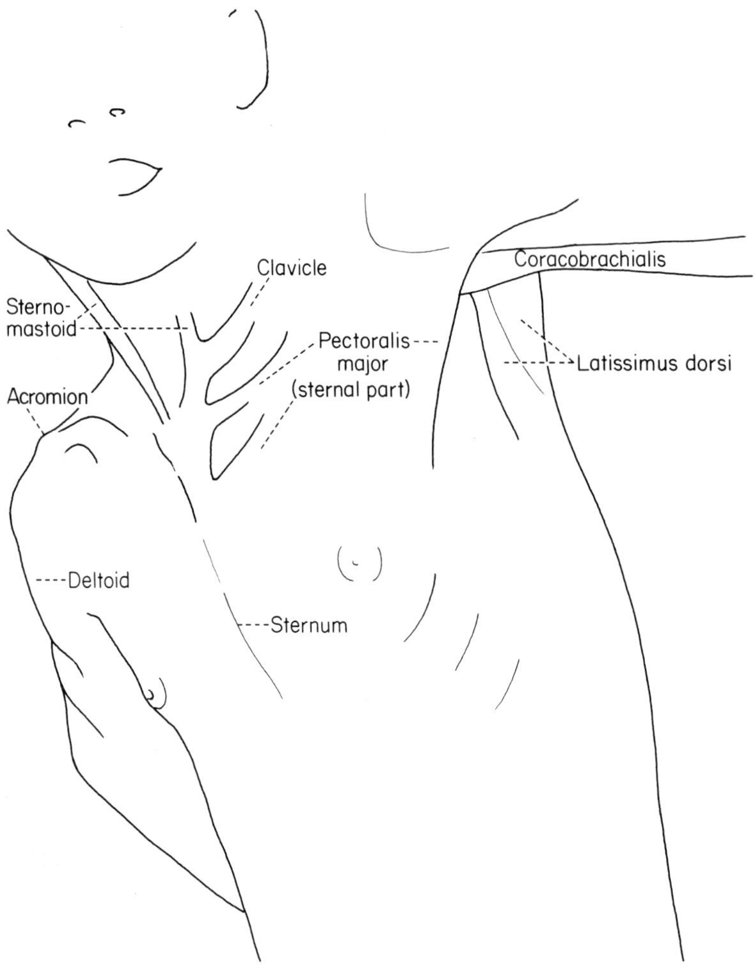

Figure 12 (♀ 11)

The pectoralis major is divided into several strands which are grouped under the sternal and the clavicular part. By pushing downward with the arm (adduction in the shoulder joint, see also Fig. 15) the sternal part becomes visible. The anterior border of the latissimus dorsi forms with the lateral border of the pectoralis major a triangular hollow, the armpit (see also Fig. 13). Contour of the right nipple and areola shows first swelling at onset of puberty.

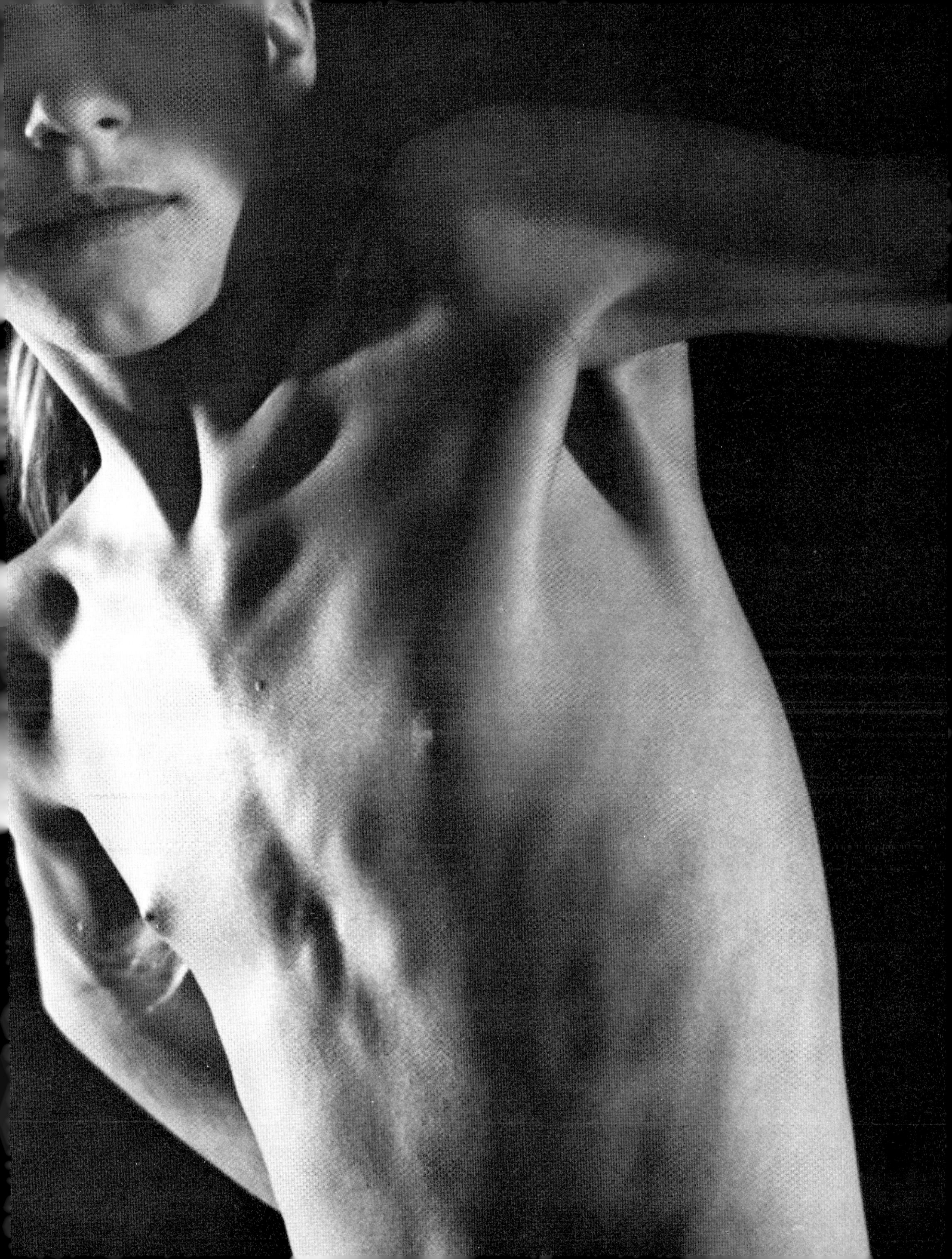

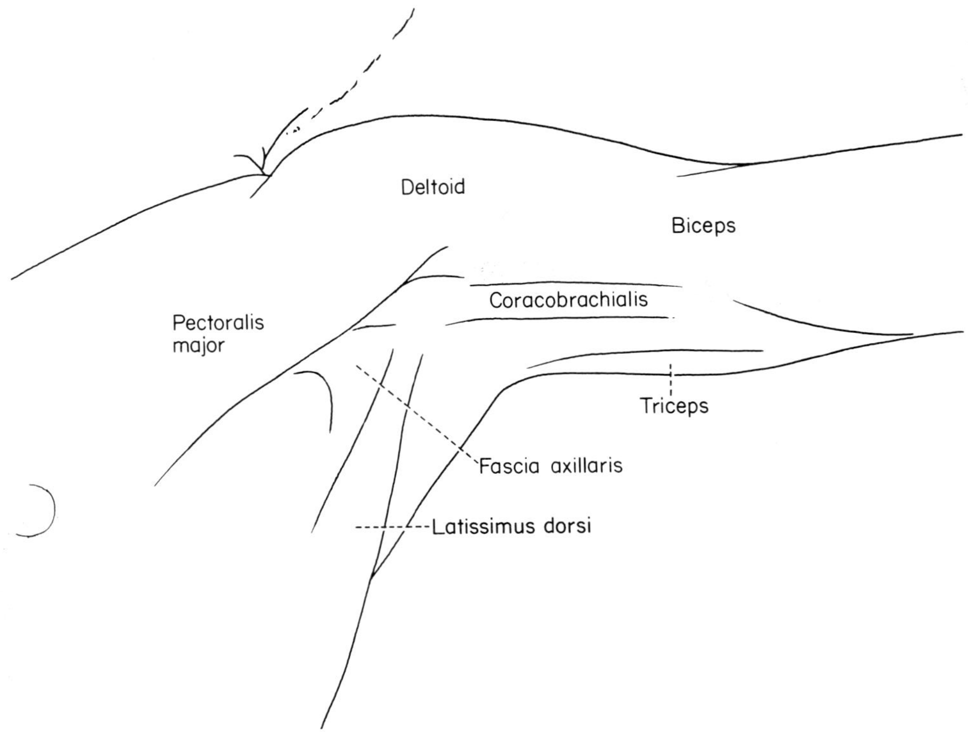

Figure 13 (♂ 9)

Adduction in the shoulder joint.

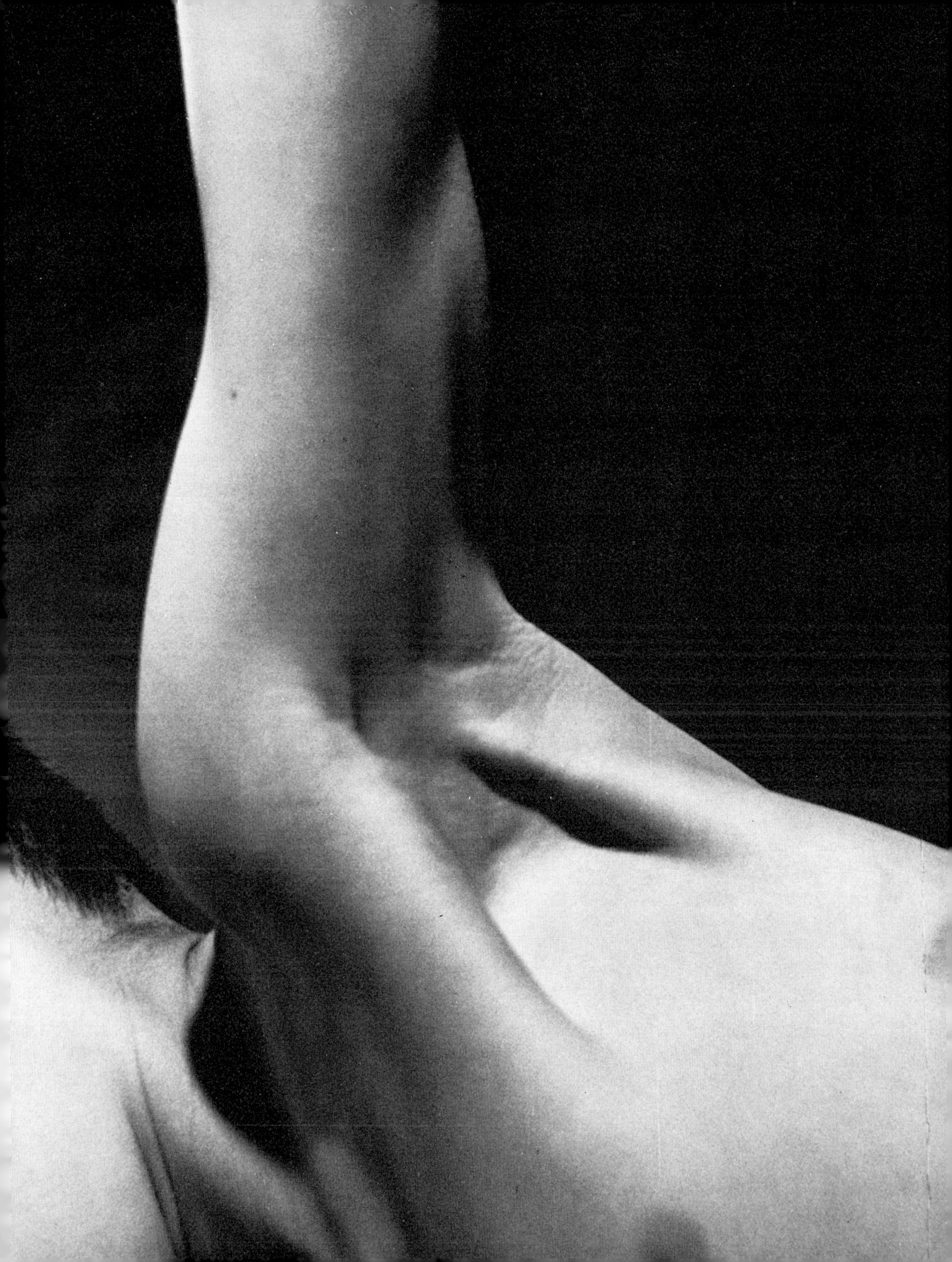

Figure 14 (♀ 44, ♂ 53, ♂ 72)

By hooking the fingers of both hands above the head and then pulling against each other many muscles around the shoulder joint become clearly visible, even where skin has lost some resilience with age. All three subjects had participated in physical fitness classes (two to three times per week) since their childhood.

Figure 15 (♂ 8)

In the upper figure the subject is pushing statically downward with the hand against the resistance of a stick; in the lower figure he is pulling upward on the stick. This demonstration brings out the separate parts of the pectoralis major. The clavicular part contracts for the upward pull, while the sternal part contracts for the downward push.

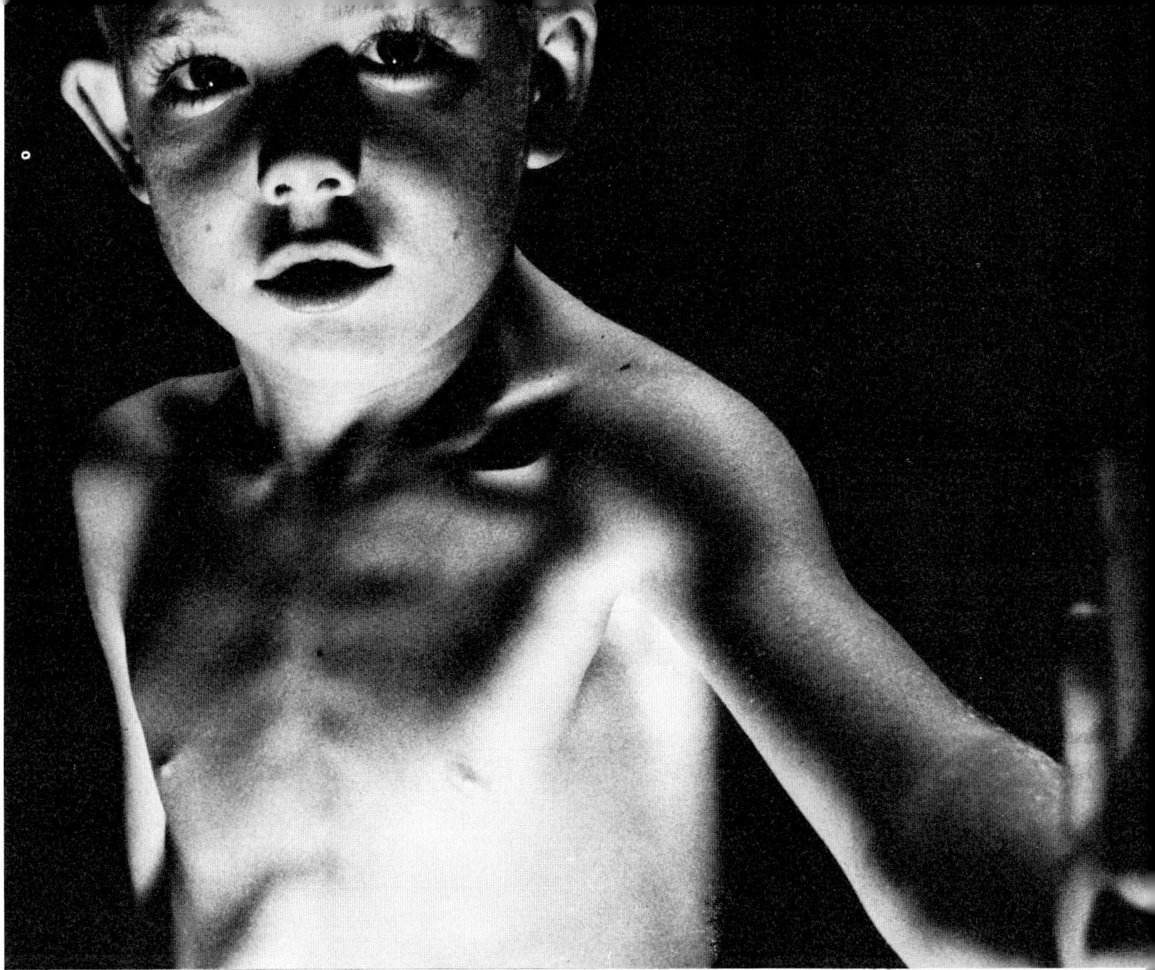

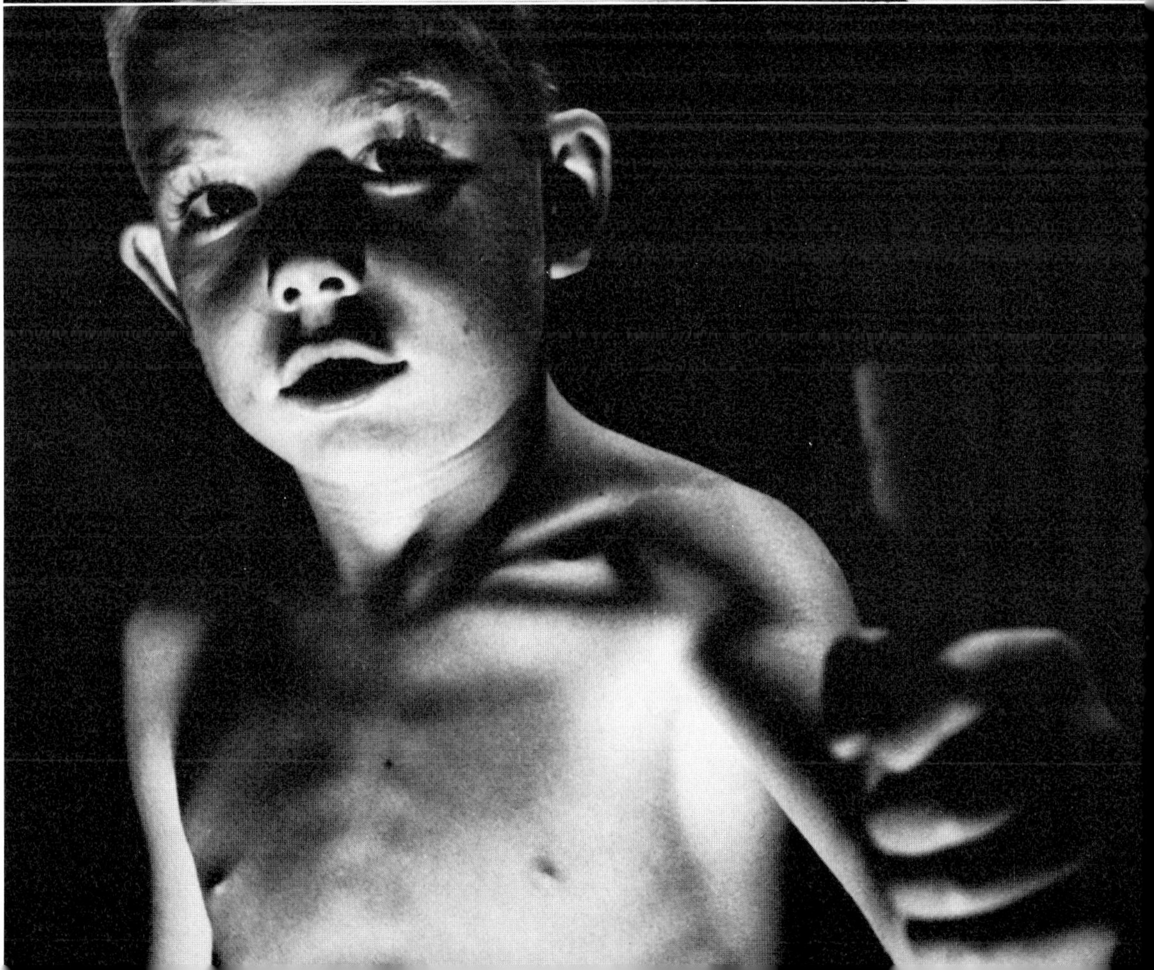

Figure 16 (♂ 6)

The boy attempts to push the ends of the stick together (static horizontal flexion-adduction in the shoulder joint) to demonstrate the separate strands of the pectoralis major. The insertion end of this muscle disappears between the deltoid and the biceps to attach to the upper part of the humerus.

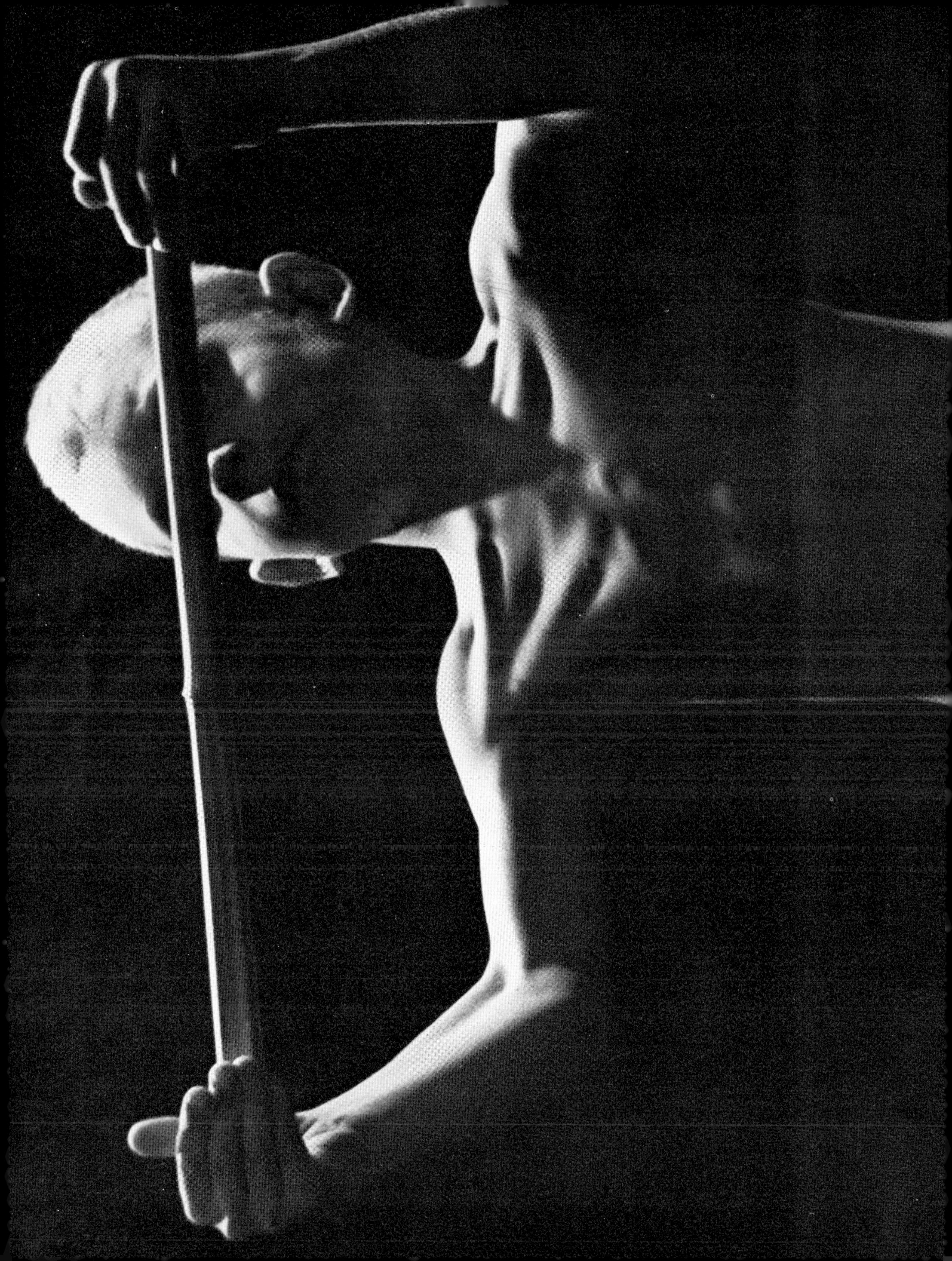

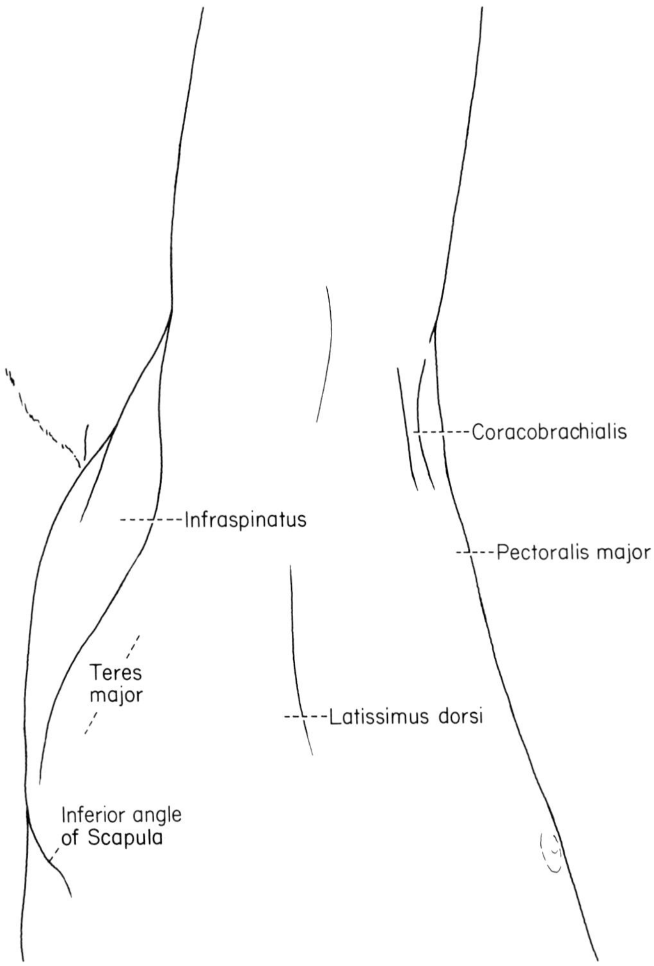

Figure 17 (♂ 9)

The hands are pulling against each other above the head while the elbows are brought backward. Attachment of the teres major to the inferior angle of the scapula is visible.

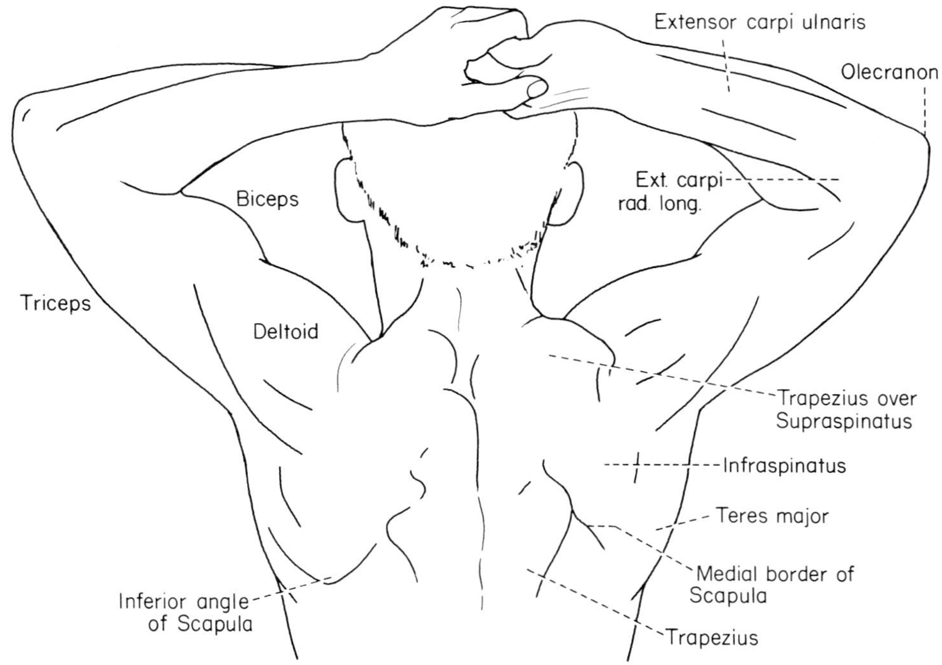

Figure 18 (♂ 21)

Many years of gymnastics training have developed the shoulders and upper back in this subject. The trapezius is thick and well defined (see also Fig. 13). Together with the supraspinatus it forms a bulge medial-ward of the deltoids when the arms are raised and the shoulder blades pulled together (adducted).

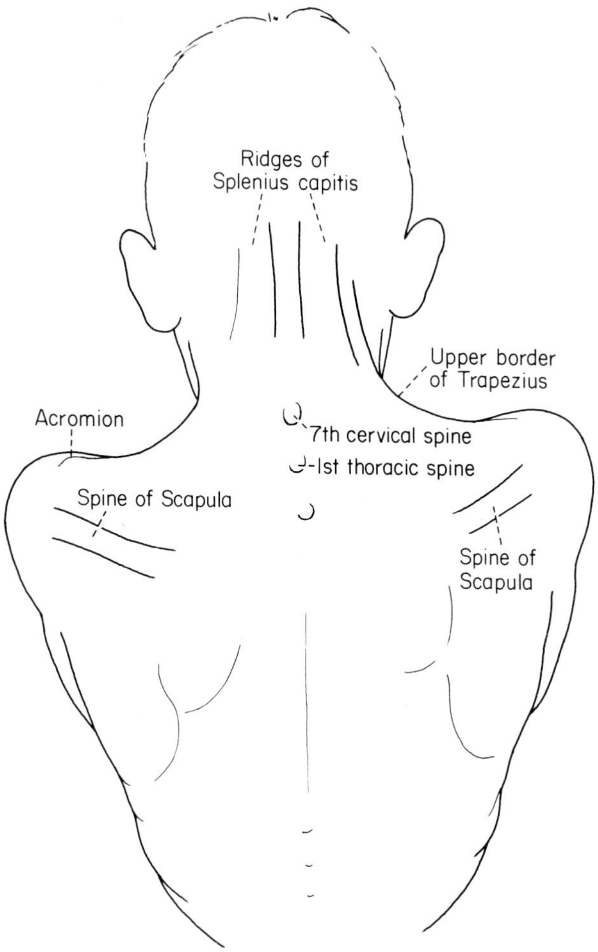

Figure 19 (♂ 8)

The upper trapezius in children is thin enough to show the underlying splenius capitas. The latter muscles are easily demonstrated by having the subject extend the neck (lift the head) against resistance (gravity).

By bending the subject's head forward the seventh cervical spine becomes the most prominent one and can thus be found easily even in fat subjects.

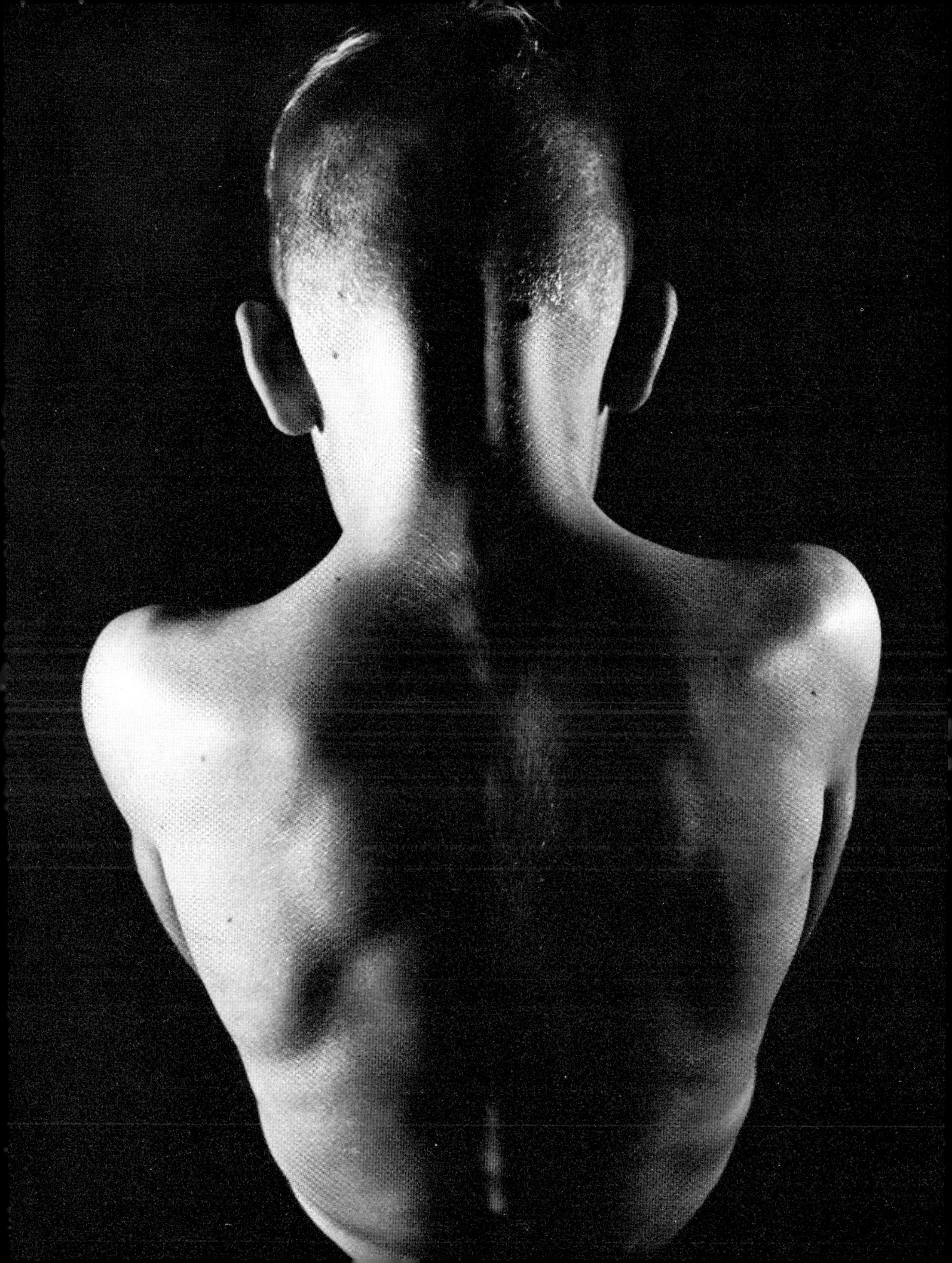

Figure 20 (♀ 17)

The subject is attempting to lift a heavy weight with the right hand. The supporting action of the trapezius (part two) muscle can be seen as a rather steeply sloping ridge on the shoulder. At rest the steepness is usually less in females than in males. Muscular and skeletal contours generally are smoother in adult females because of overlying tissues (breast, adipose tissue).

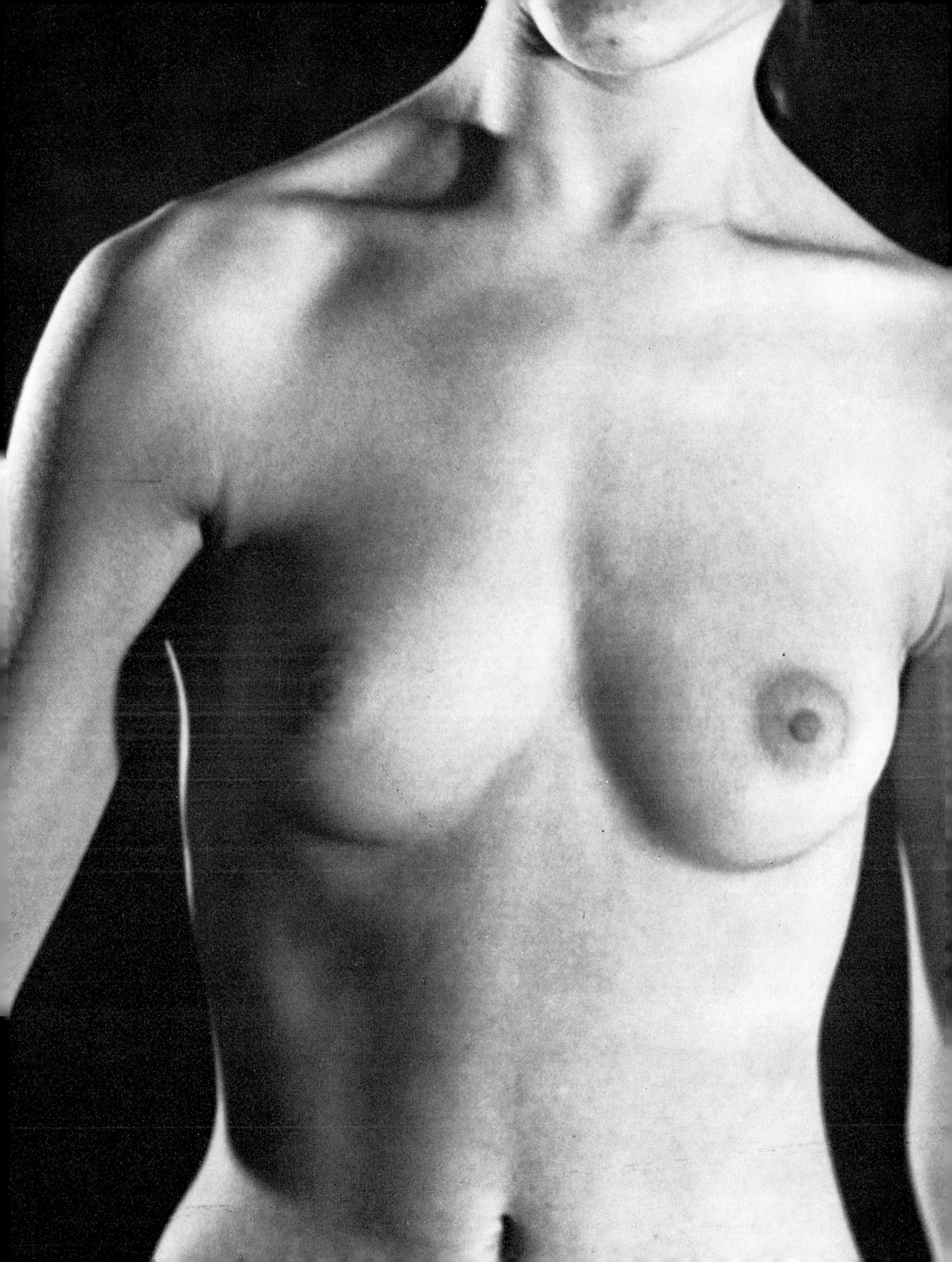

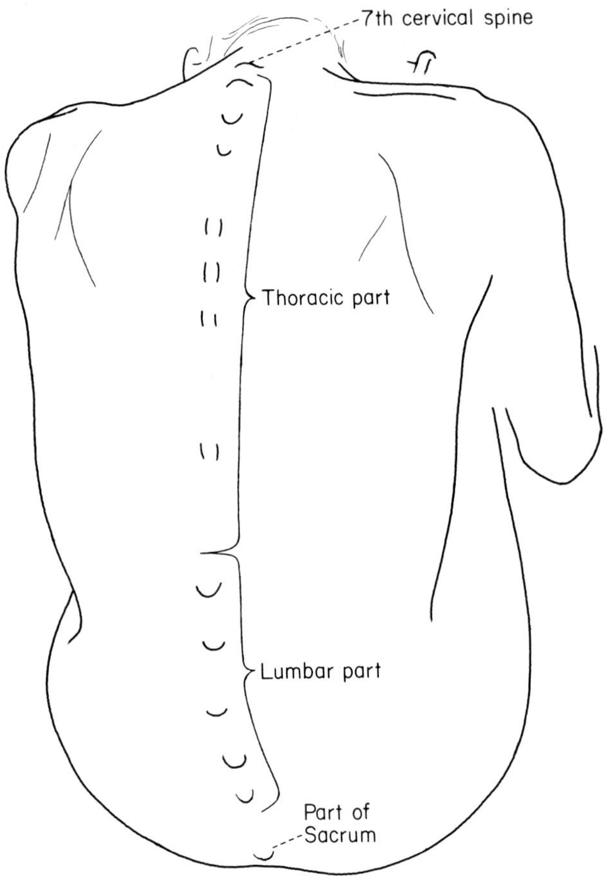

Figure 21 (♀ 17)

Extreme flexion of the neck and trunk brings out the spinous processes of the spinal column. By counting from the seventh cervical (uppermost large protuberance) downward we can identify the twelve thoracic vertebrae, the five lumbar vertebrae, the sacrum and coccyx.

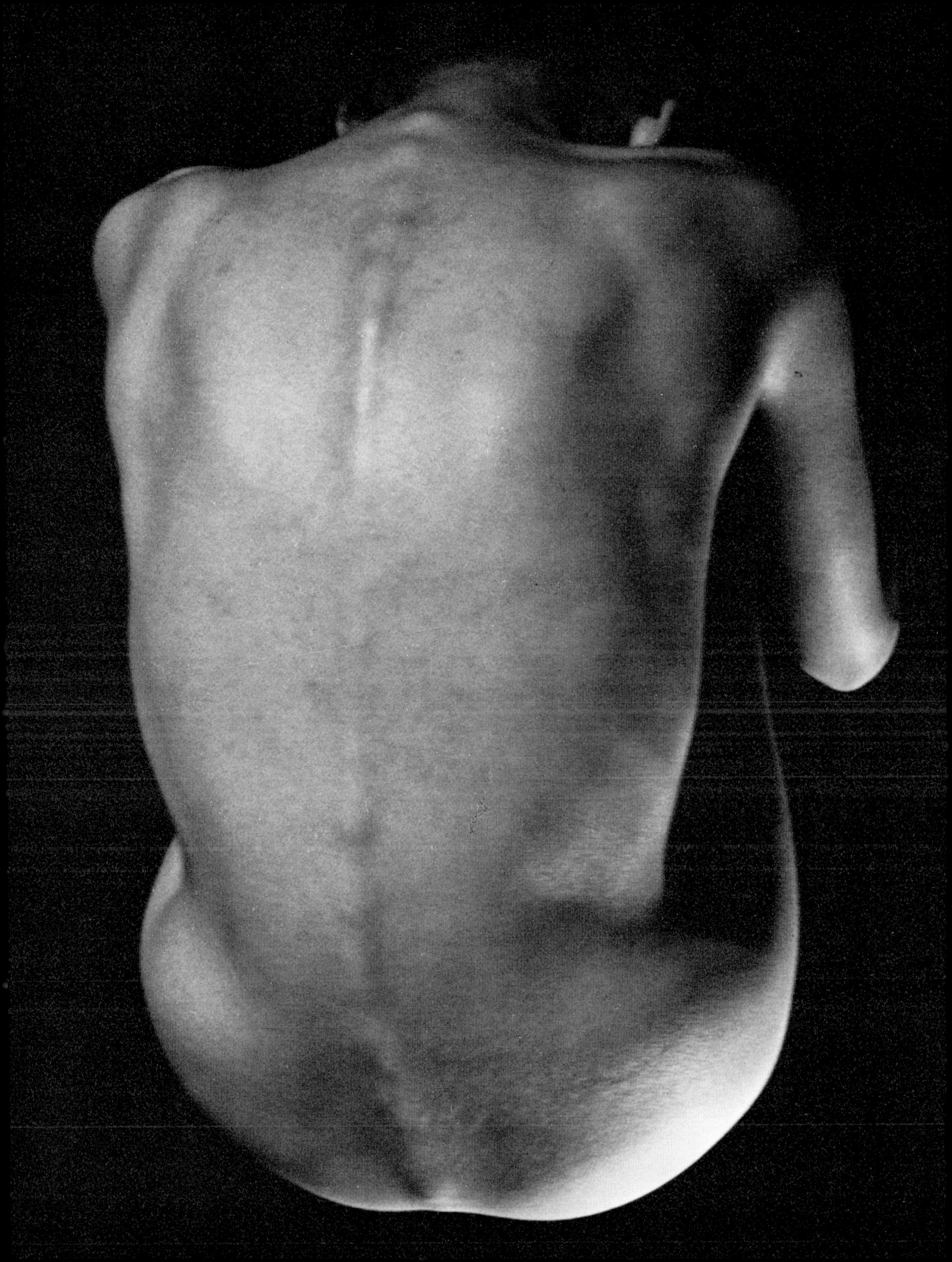

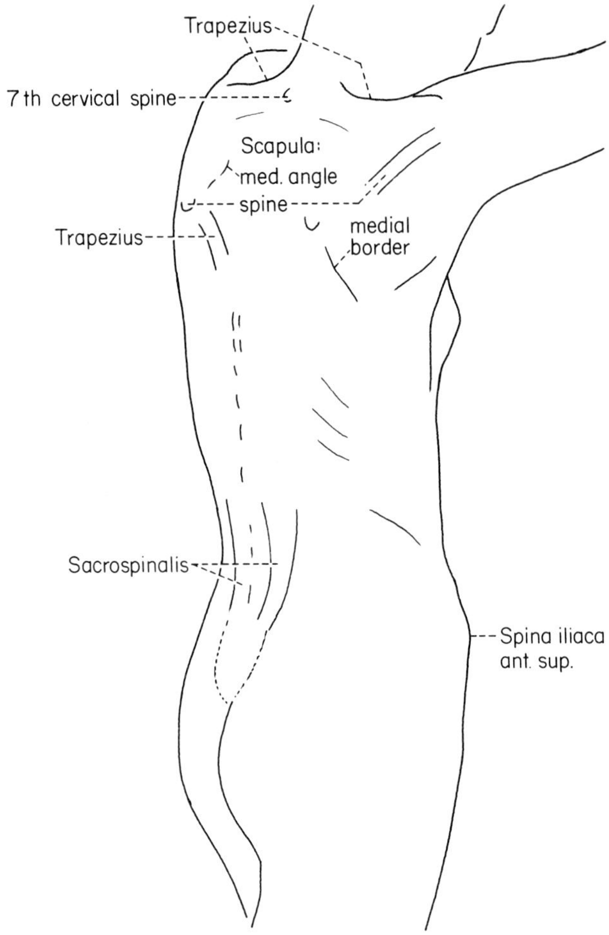

Figure 22 (♀ 15, late maturation, tall thin)

In thin subjects the rotation of the scapula can be observed when the arms are raised above horizontal. The sacrospinalis muscle group is best seen in the lumbar region where the left and right bundles will pop up alternately in walking. (Compare with Figs. 26 and 27).

Figure 23 (♀ 15)

Same subject as in Figure 22 from a slightly different angle.

Figure 24 (♀ 12)

The sacrospinalis muscles are visible even in well-filled subjects. So are the locations of the posterior superior spines of the ilium which are indicated by the two dimples at the side of the lowest part of the lumbar region. The imaginary diamond-shaped area, formed by lines drawn from the first lumbar spine to the dimples and from there converging to the uppermost point of the buttock fold, has significance in the detection of unsymmetrical walking patterns (see also Fig. 28).

Figure 25 (♀ 11)

The great mobility of the scapulae can be demonstrated by bringing the elbows back while the upper arm rotates inward. (Try to scratch your scapula from below with your thumbnail.) The inferior angles and medial borders of the scapulae then come away from the rib cage.

The dimples, representing the location of the posterior superior spine of the ilium, are clearly visible.

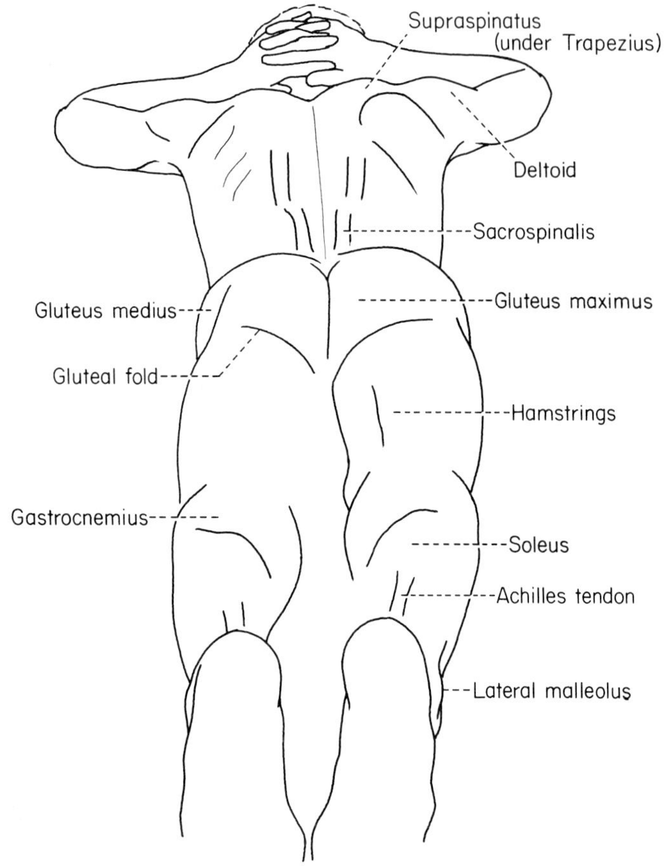

Figure 26 (♂ 26)

When the subject arches his body upward while lying in a prone position, all the posterior muscle groups of the body come into action against the resistance of gravity.

Figure 27 (♂ **26)**

The same subject as shown in Figure 26 from a different angle.

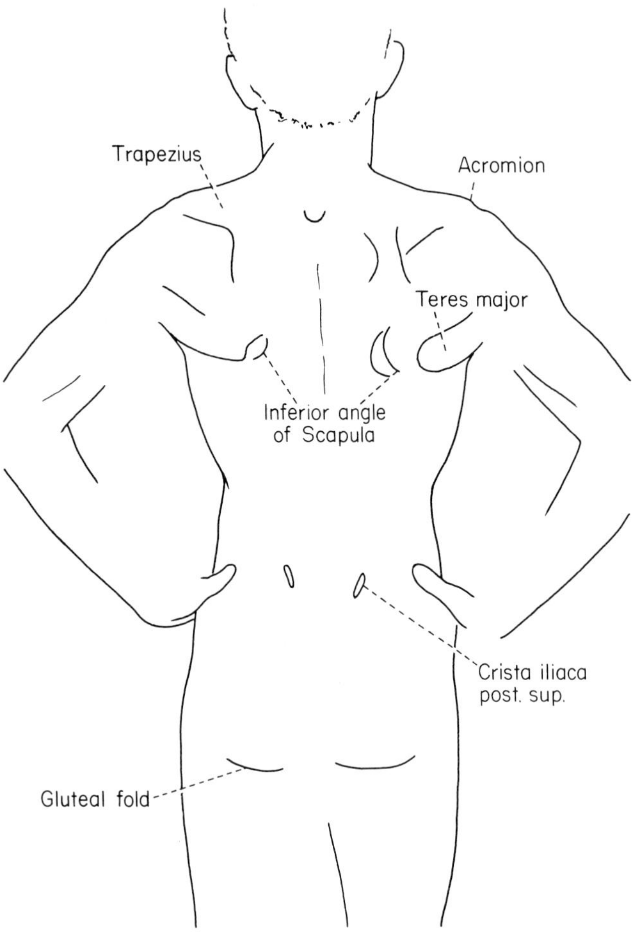

Figure 28 (♂ 23)

The subject attempts to lower the shoulders against the resistance provided by pushing the hands down on the hips. This action provides a clear view of the teres major muscles.

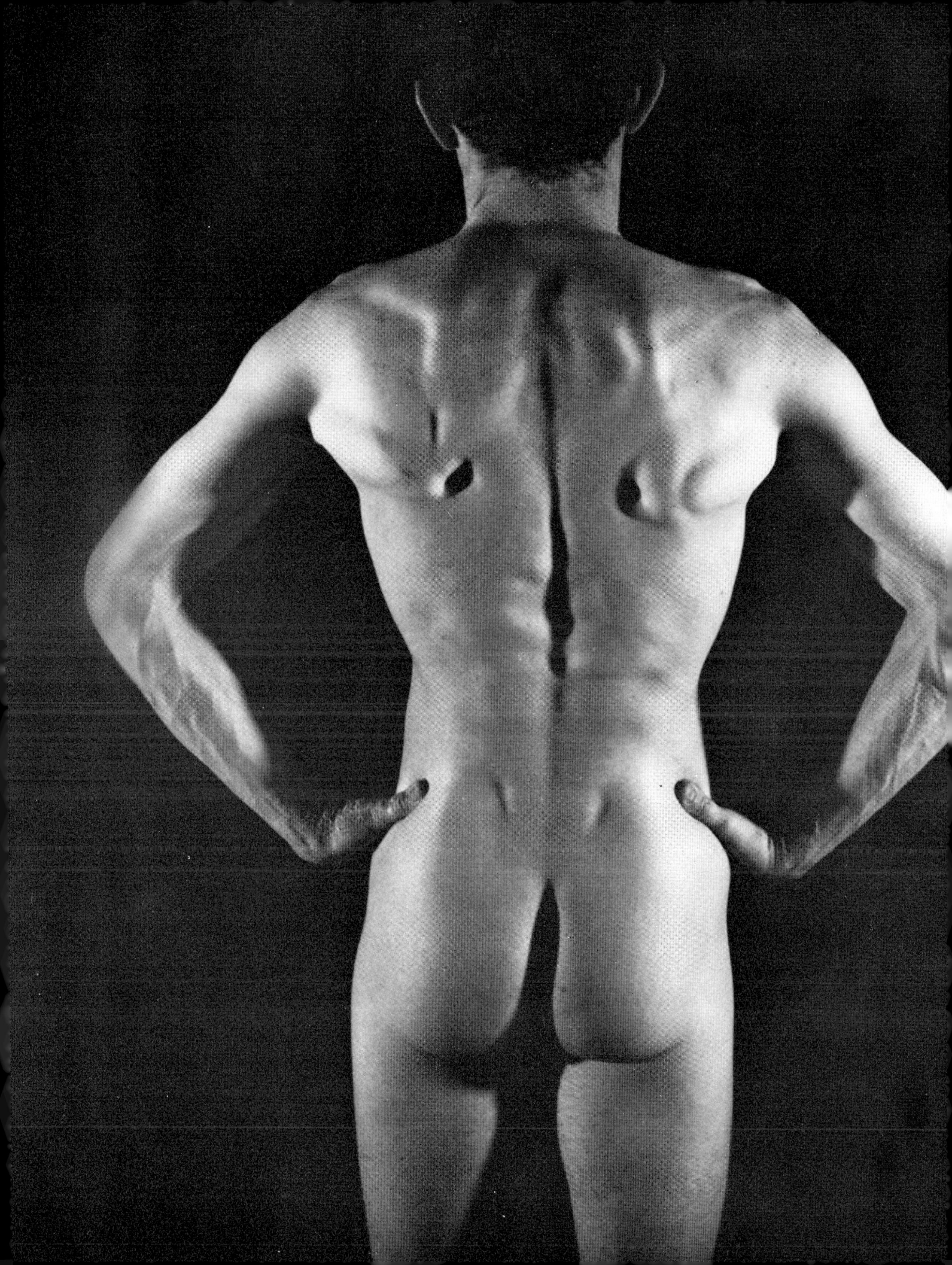

Figure 29 (♀ 15)

The same subject as in Figures 22 and 23. The curvature of the spine in flexion is shown in this thin subject. The lumbar part and the upper thoracic part curve most. The ribs are visible on the back. The bulging digitations of the lower part of the serratus anterior are visible on the anterior part of the trunk. Behind the right knee the tendon of the biceps femoris stands out.

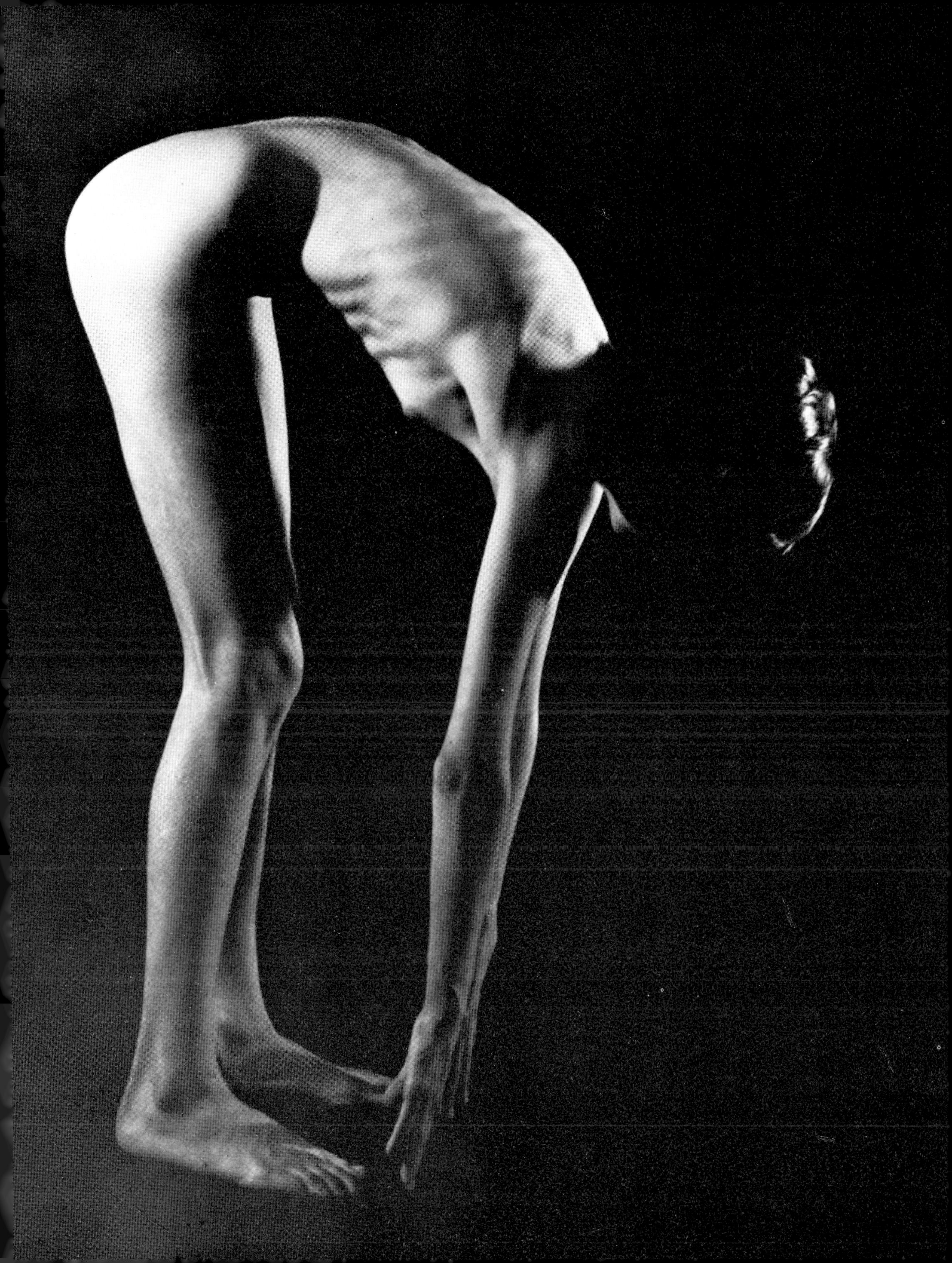

Figure 30 (♀ 23)

Fat deposits are not evenly distributed. The thickness of the layer of adipose tissue varies from place to place. Deposits are still slight on the lower thoracic and lumbar part of the spine. The deposits on the spina iliaca posterior superior were slight. In contrast to some other subjects (Fig. 121) the fat in this individual was hard and not flabby, which may point towards a wide variety of adipose tissue composition in fat persons.

Figure 31 (♀ 23)

The same subject as shown in Figure 24 attempts to "suck in" the abdomen. Fat deposits on the abdomen and above the gluteus maximus are visible.

Figure 32 (♀ 17)

Extension in the spine occurs for the most part in the lumbar region.

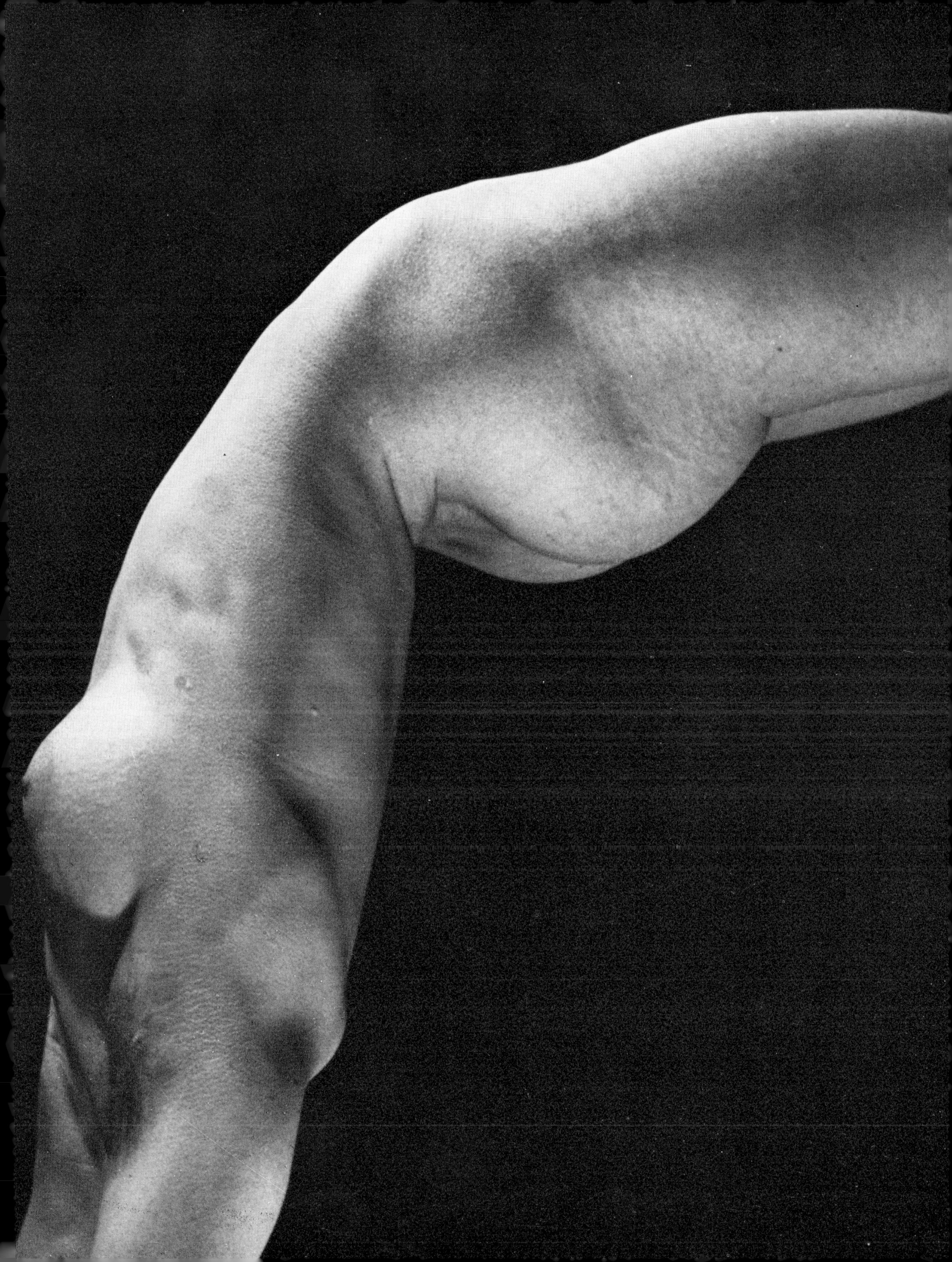

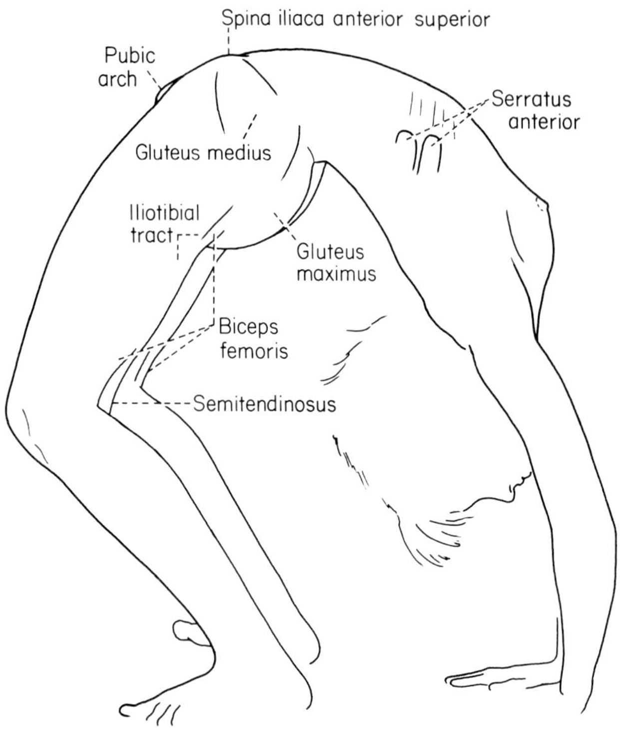

Figure 33 (♀ 16)

The greatest mobility for the extension in the spine is in the lumbar and in the cervical region, the thoracic part is rather stiff.

In order to reach the greatest backward arch it is necessary to tilt the pelvis as far back as possible. A strong ligament in the groin region (iliofemoral ligament) prevents further tilting of the pelvis. When the hip joints are in an outward rotated position, however, slightly more backward tilt is possible. The knees have to be flexed to provide some extra "bend." Stiffness in the shoulders can be compensated for by bending the elbow.

The present subject also increased the forward mobility of the ankle by placing the feet in an abducted and everted position (pronation).

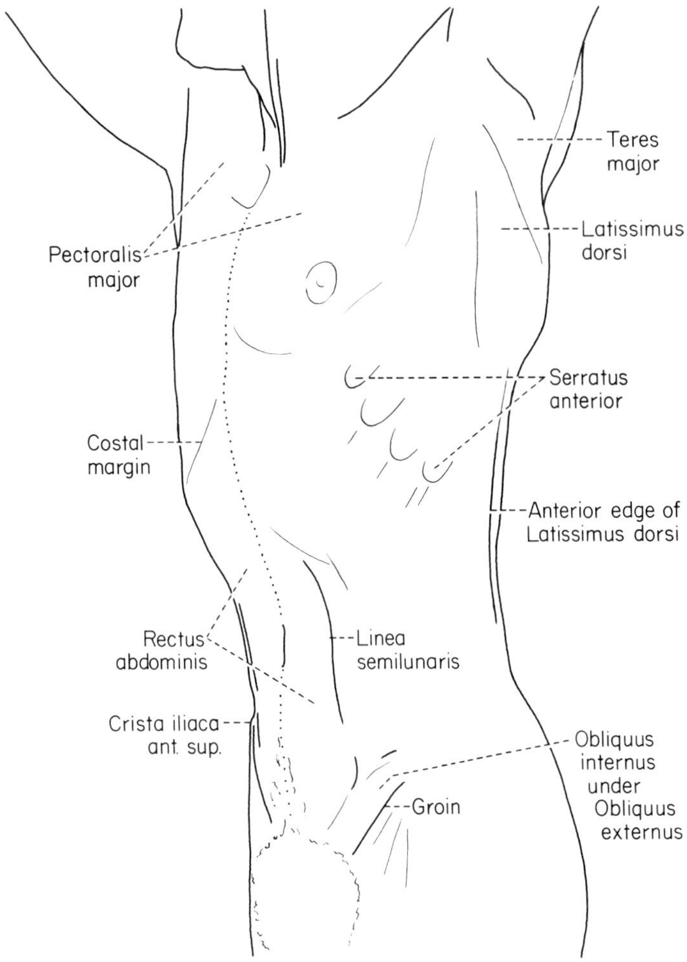

Figure 34 (♂ 21)

The chest is raised by a deep inspiration and the abdominal wall is "sucked in." This action demonstrates the position of the linea semilunaris which indicates the lateral edges of the rectus abdominis. The vertical furrow in the center of the abdominal wall indicates the linea alba.

The slight bulge of the internal oblique muscle can be seen through the overlying fascia of the external oblique.

Pubic hair in the male has a greater tendency to cover the linea alba upwards from the pubis than in females in whom the hairline is more often horizontal.

The lower edge of the chest, the costal margin, is formed by the cartilaginous part of the seventh to tenth ribs. It starts at the sternum just above the joint with the xiphoid process, and sweeps downward and lateral-ward to only a few inches from the upper crest of the ilium. There exists a wide variety in the size of the angle between the costal margins.

Figure 35 (♂ 18)

The cartilaginous part of the seventh to ninth ribs in this subject flared outwards in a forced maximal inspiration.

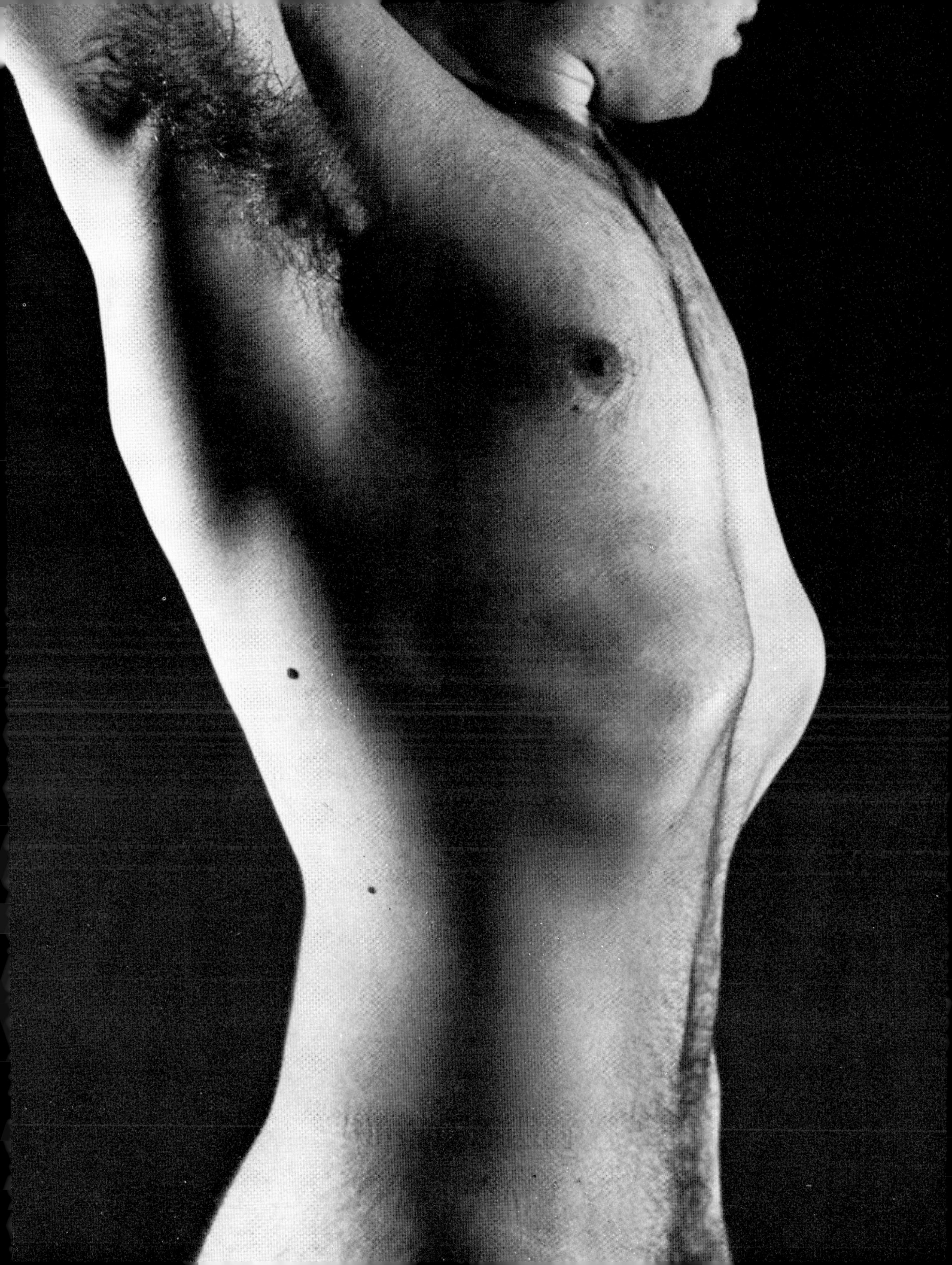

Figure 36 (♀ 12)

The mobility of the ribs is shown by this subject who first extended the thoracic part of the spine to the extreme and then, without changing the position of the head, flexed the thoracic spine. This action flattened the chest considerably.

The abdominal contents in the second action pushed the relaxed abdominal wall outward. Some adipose tissue deposit can be seen around the umbilicus.

The areolae have the characteristics of the first swelling at the onset of puberty.

Figure 37 (♀ 26)

A mother of an eight-year-old child. The breast tissue extends from the upper thorax (approximately second rib) and covers the lower part of the pectoralis major. Laterally the breast extends to the axilla. The areolae have retained the dark-brown color after pregnancy.

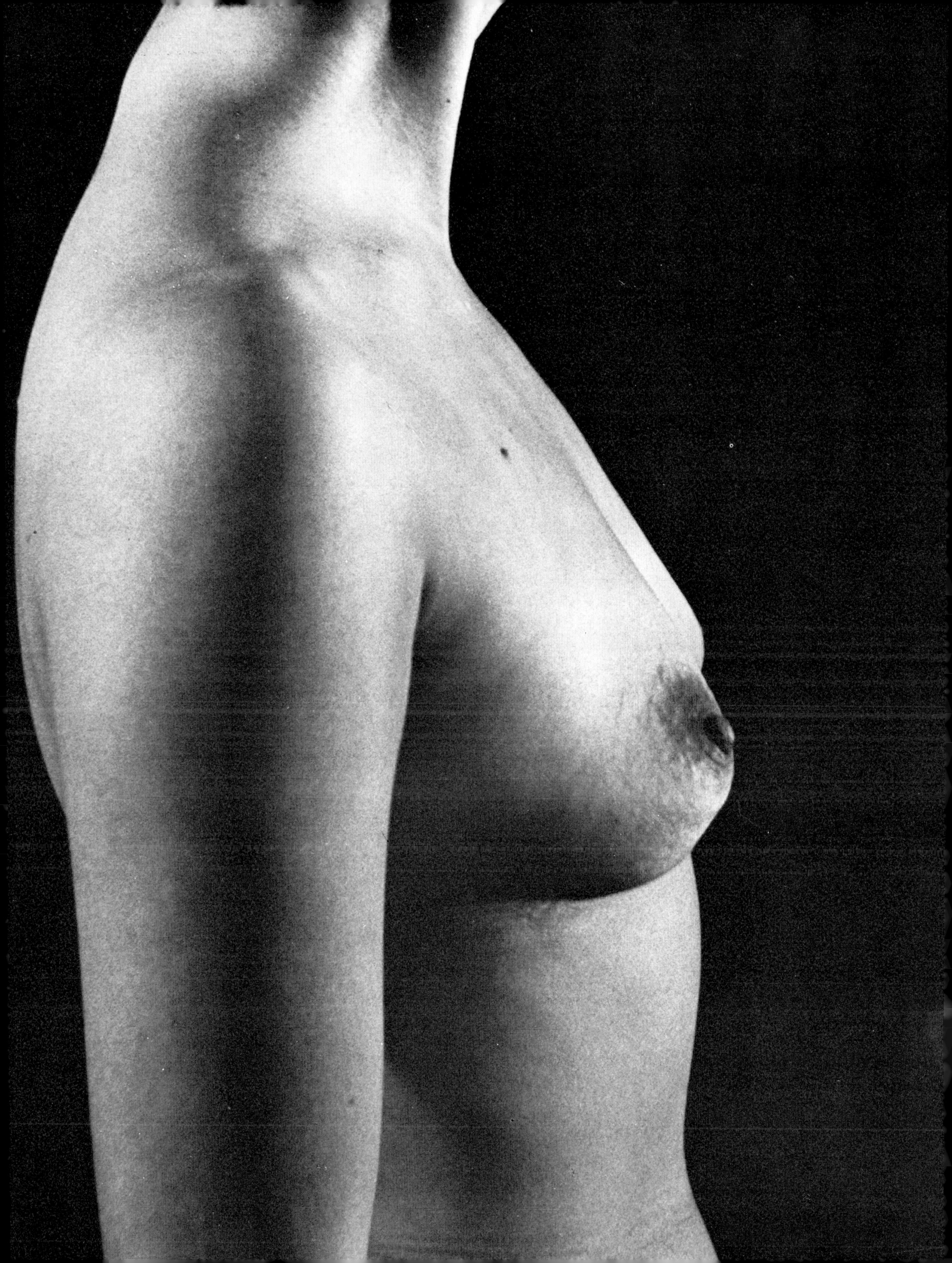

Figure 38 (♀ 34)

The venous pattern of lactating breasts (five days postpartum) shows up on infrared-sensitive film.

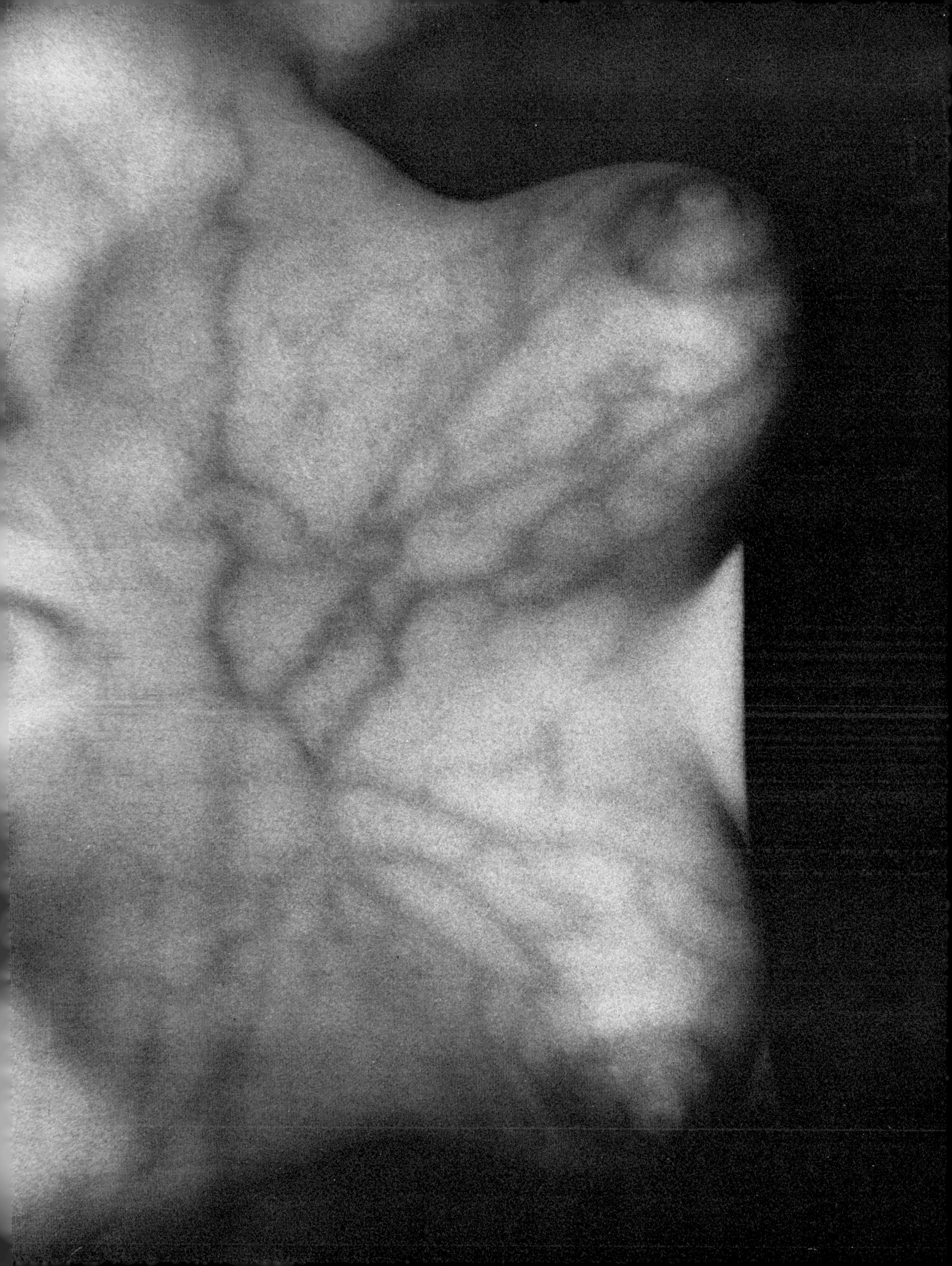

Figure 39 (♀ 22)

Virginal breast shows lightly pigmented areolae. Upwards the breast fuses its contours into the lateral margin of the pectoralis major. The subject has heavy deposits of adipose tissue which efface the bone and muscle definition, and fill up the axillary space. (Compare with Figs. 14, 17, 36.)

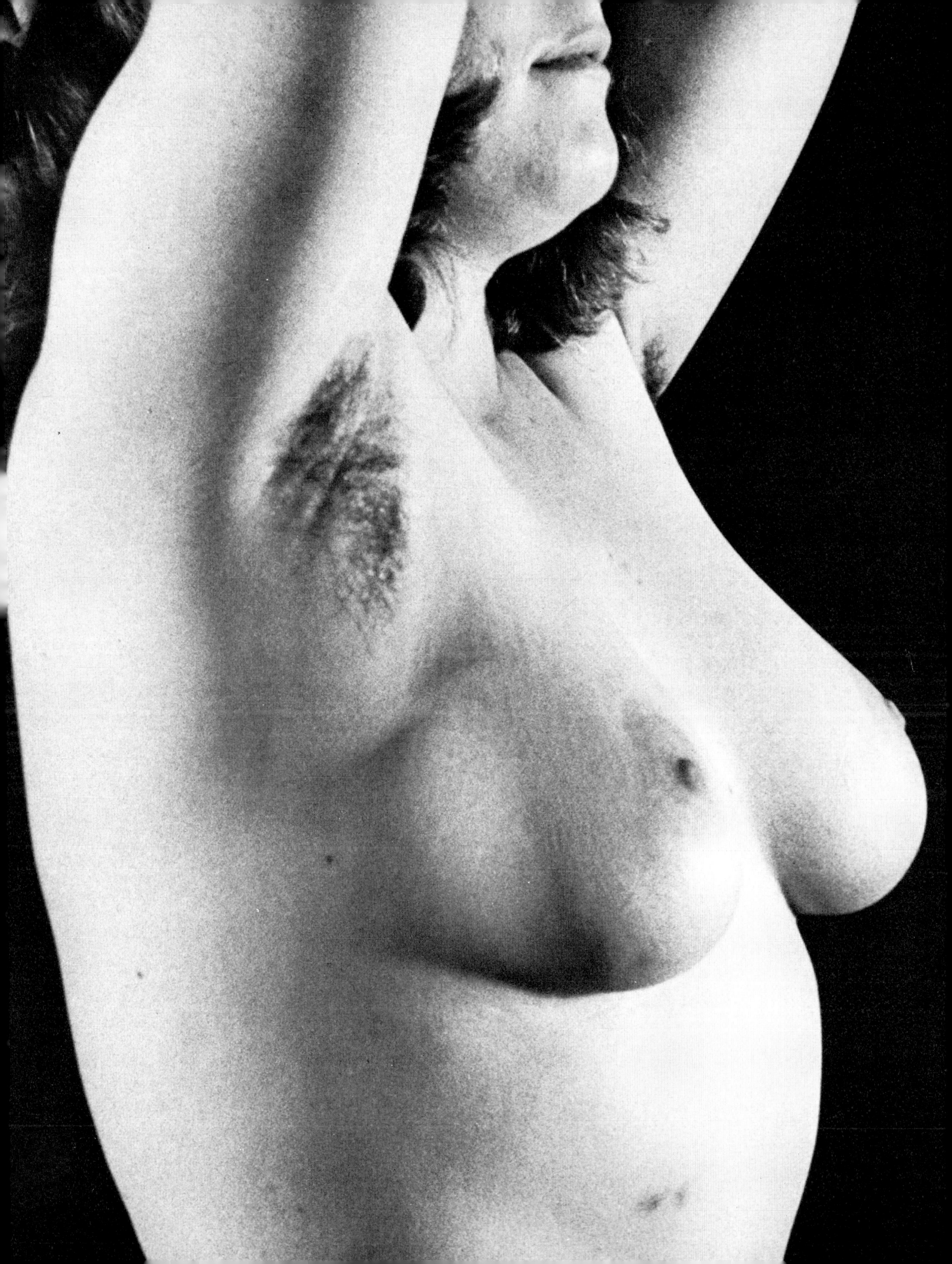

Figure 40 (♀ 26)

The curvature of the ribcage causes the breasts to point outward rather than forward. Extreme foreshortening in this figure accentuates the fact that the greatest width is found below the hip joint in females. This is due to fat deposits. (See also Figs. 68-71.) The greater pelvic width in females causes a slightly knock-kneed position of the leg when the knees are held closely together.

The patella is obscured by surrounding fat pads.

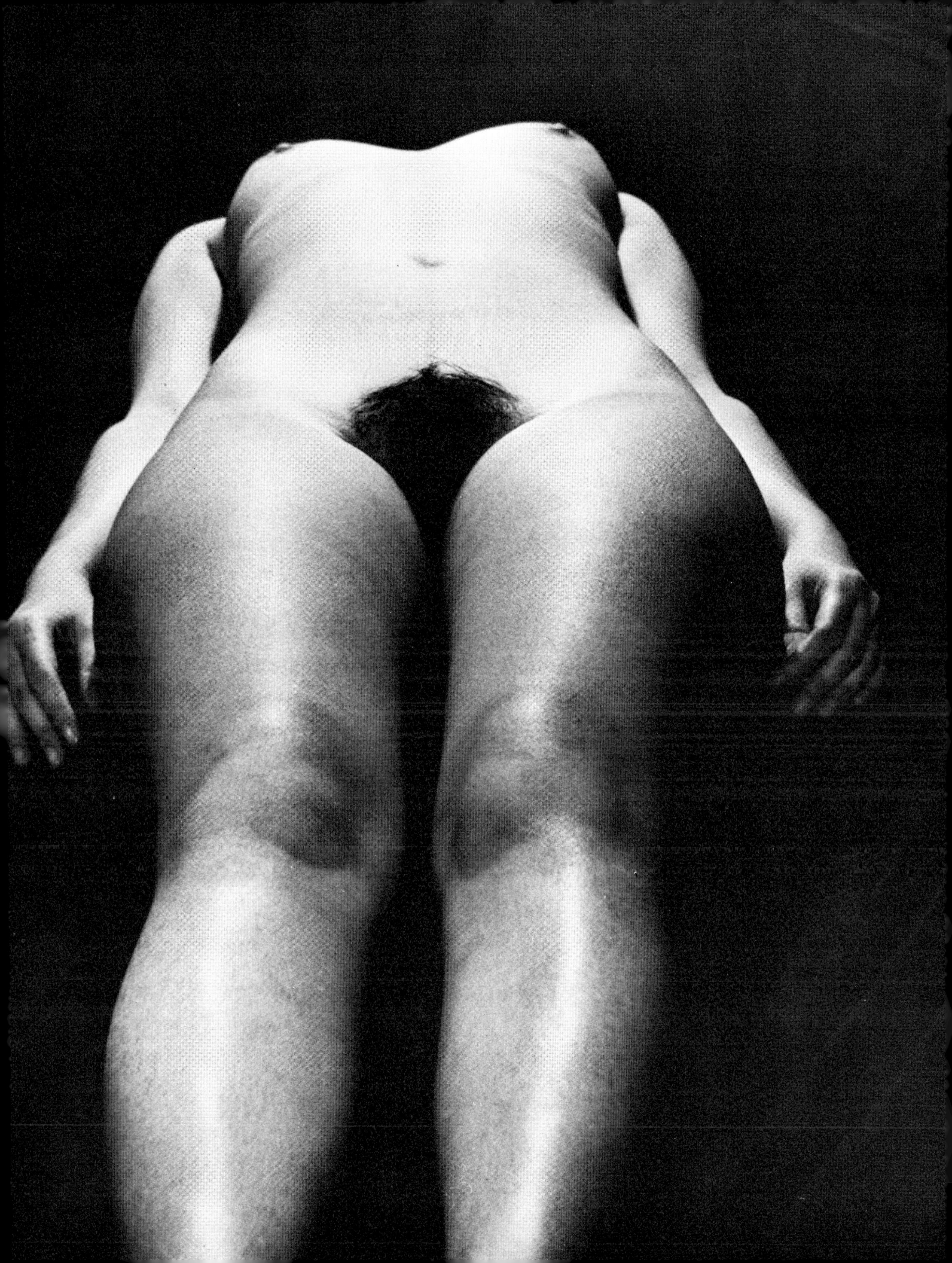

Figure 41 (♂ 28)

This subject shows abdominal muscles which were trained daily by arching backward over the edge of a table as a starting position for doing fast sit-ups. The inscriptio tendinea of the rectus abdominus divides this muscle horizontally into three distinct muscle bellies above the umbilicus. The linea alba divides the muscle lengthwise (vertically) into two lateral parts. The bulge over the iliac crest is formed by the well-developed obliquus internus and externus muscles. The latter interdigitates with the serratus anterior on the lower anterior ribs.

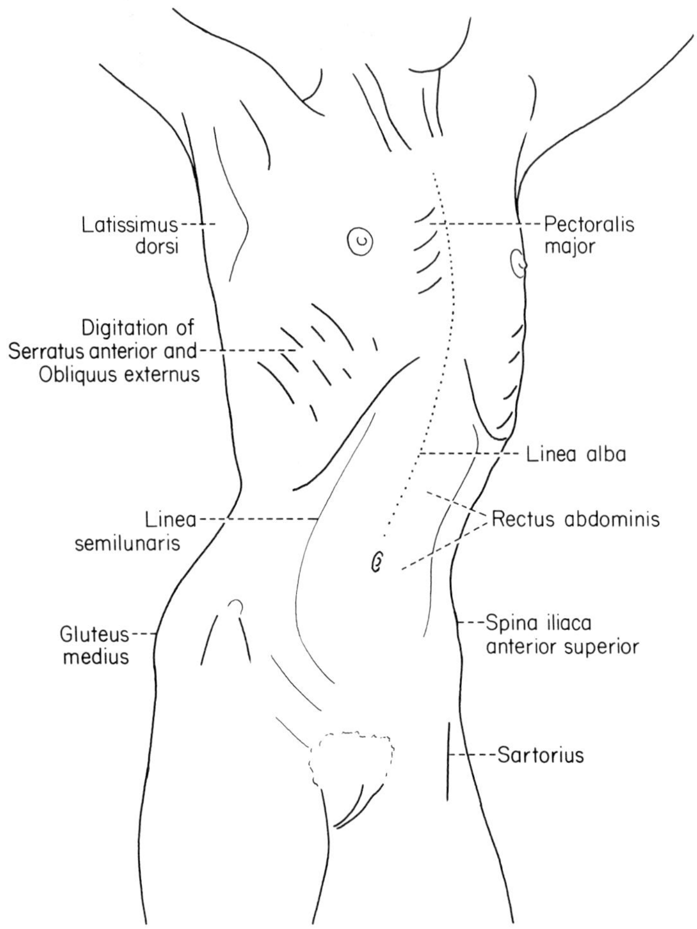

Figure 42 (♀ 13)

The subject is pulling the elbows back and pressing with the hands against the neck. With a full inspiration and a controlled flattening of the abdomen the anterior trunk muscles become clearly outlined.

Supporting the weight on one leg brings out the position of the gluteus medius.

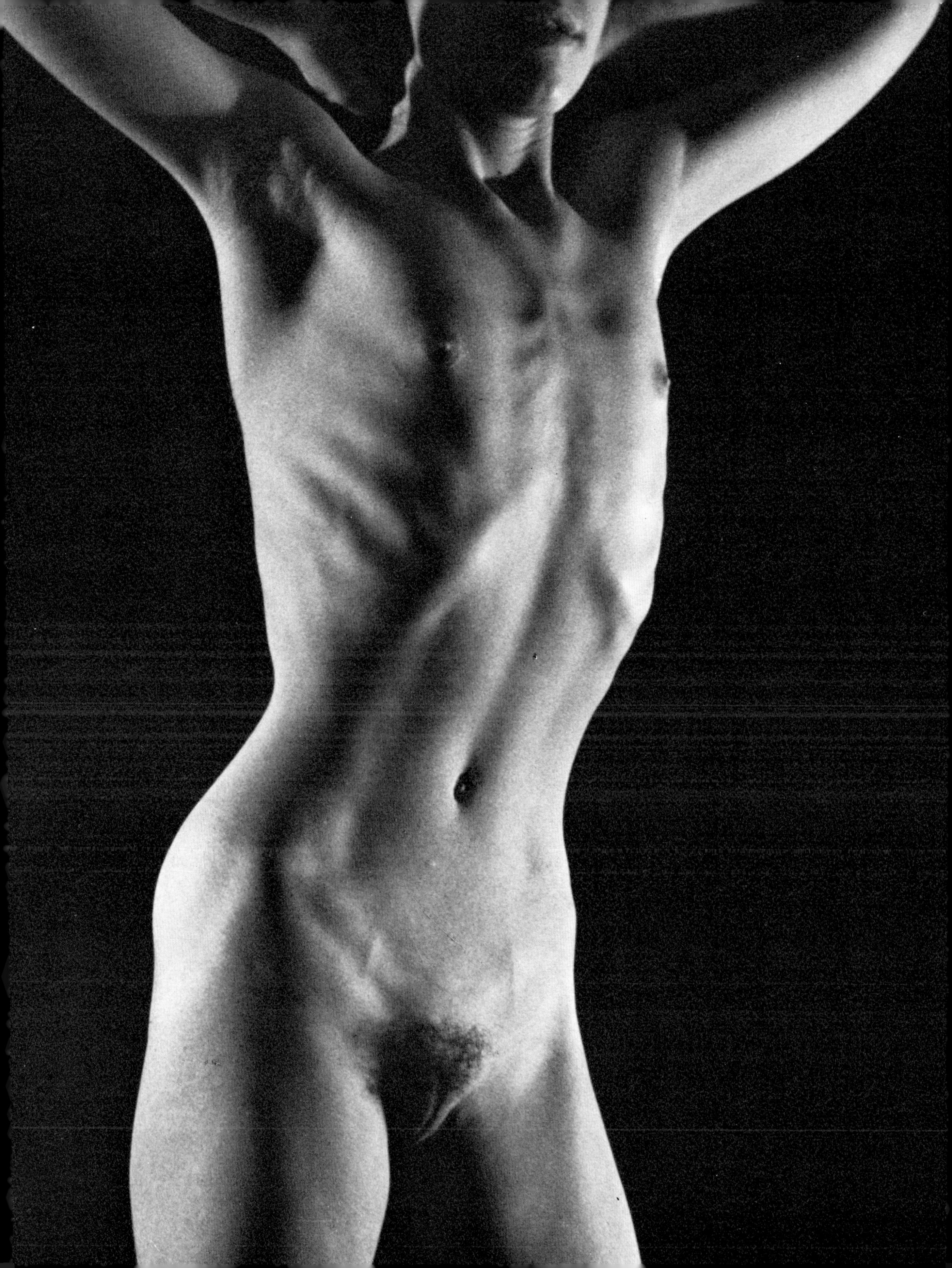

Figure 43 (♂ 29)

Muscle definition is vague in normal nonathletic adult males as contrasted with the sharp definition in a subject who trains in progressive resistive exercise (including barbell exercises) (see Fig. 41). Because of normal fatty deposits, females show more rounded contours in contrast to the angular appearance of the male.

Figure 44 (♀ 13)

The same subject as shown in Figure 42 now bends over and demonstrates the adipose tissue deposits on the abdomen. Exercise alone (she is a good ballet dancer) may not be enough to counteract the fattening process which often accompanies the onset of puberty in females.

Figure 45 (♀ 24)

In this standing posture of a six-month pregnant subject the abdominal weight tends to tilt the pelvis forward. In order to compensate for this forward shift of the center of gravity the shoulders are brought back. The result is an exaggeration of the abdominal protuberance and the lower back curvature (lordosis).

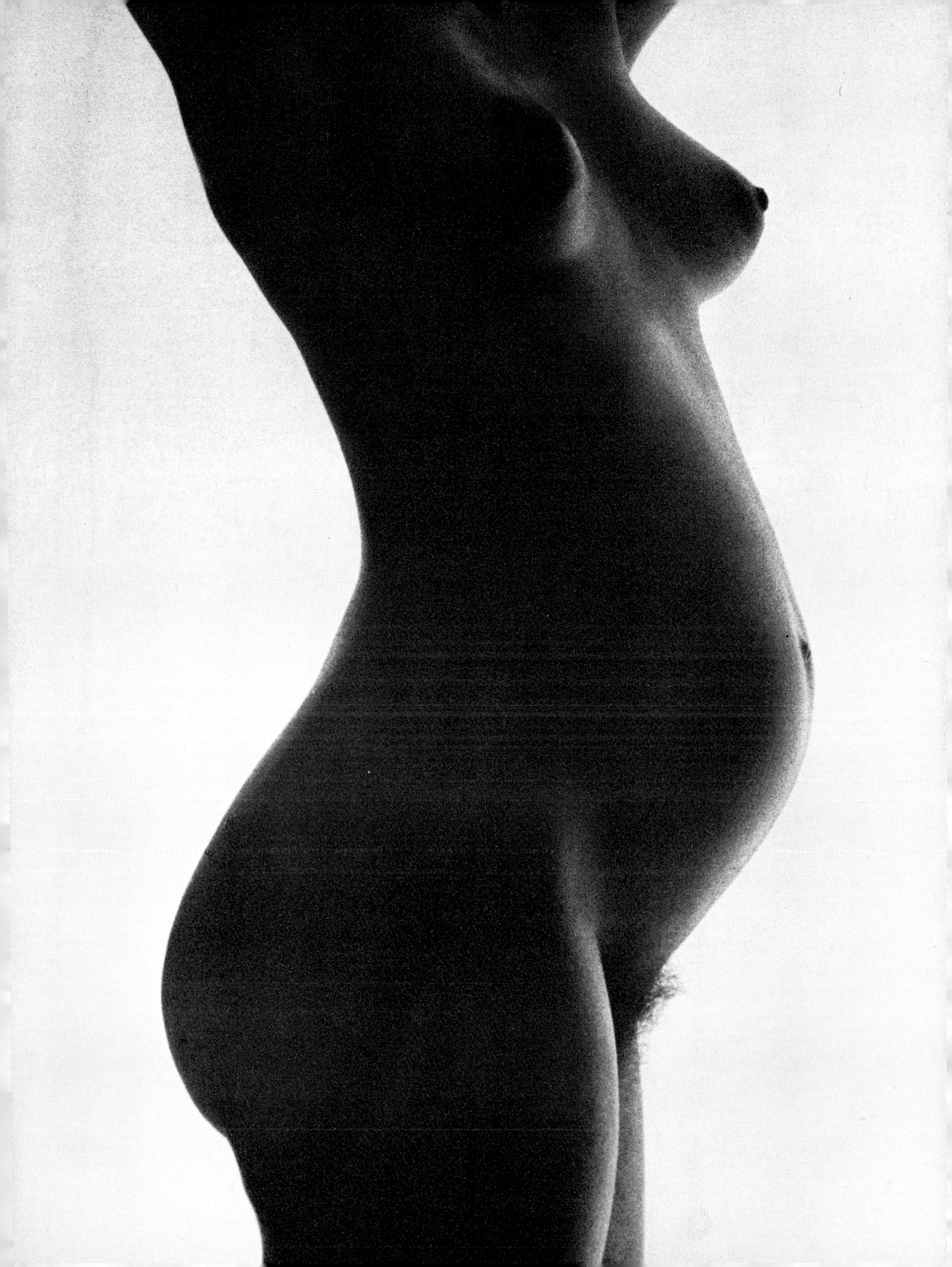

Figure 46 (♀ 8)

Normally the lateral border of the rectus abdominis is seen as two parallel furrows on the abdominal wall (compare with Figs. 42, 50). The individual shown in this figure demonstrates how the upper part of these furrows is pulled laterally. The subject was lying in a supine position, knees drawn up slightly, hands on the table, next to her hips. After exhaling completely she flattened her abdominal wall. At this moment she suddenly pressed her hands down against the table surface, producing a momentary lateral distortion of the upper part of the rectus abdominis border. (This fleeting moment was "caught" on film after twelve attempts; it was that short!)

The explanation for this distortion may be as follows: The aponeurosis of the transversalis goes around the rectus abdominis and also forms part of the lumbodorsal fasciae of the latissimus dorsi. When the girl pushed her hands down against the table the tension on the aponeurosis pulled the parallel furrows outwards. The lower part of these furrows were already pulled medialward by the transversalis which acts in forced expiration, causing the inferior attachment of the rectus abdominis to appear more V-shaped. The latter shape at the inferior end may be accentuated by the narrowness of the pelvis in a child. Also, the lowest segment of the rectus abdominis is not developed very strongly in a child of this age and this may result in a V-shaped compression by the aponeurotic sheath.

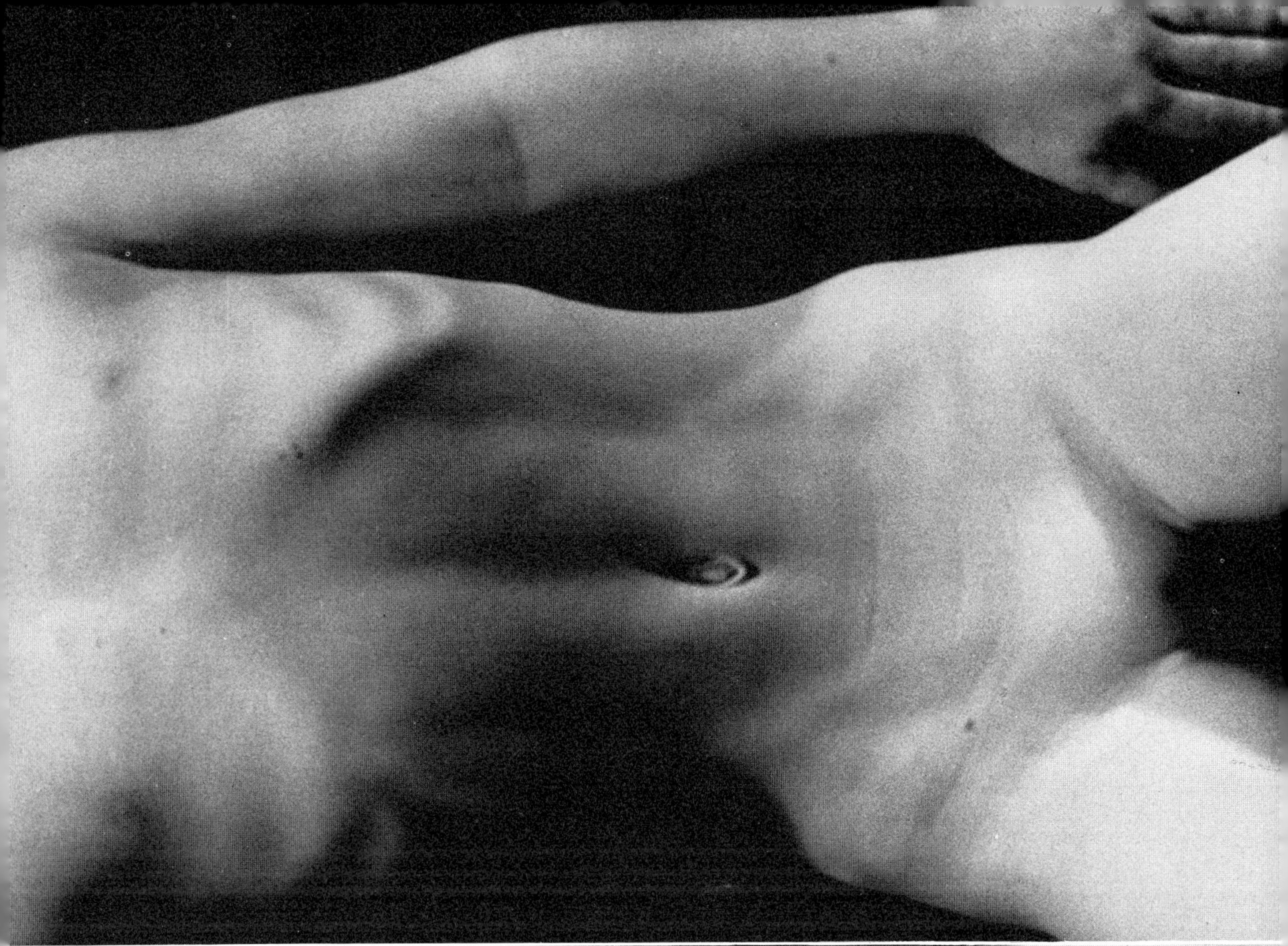

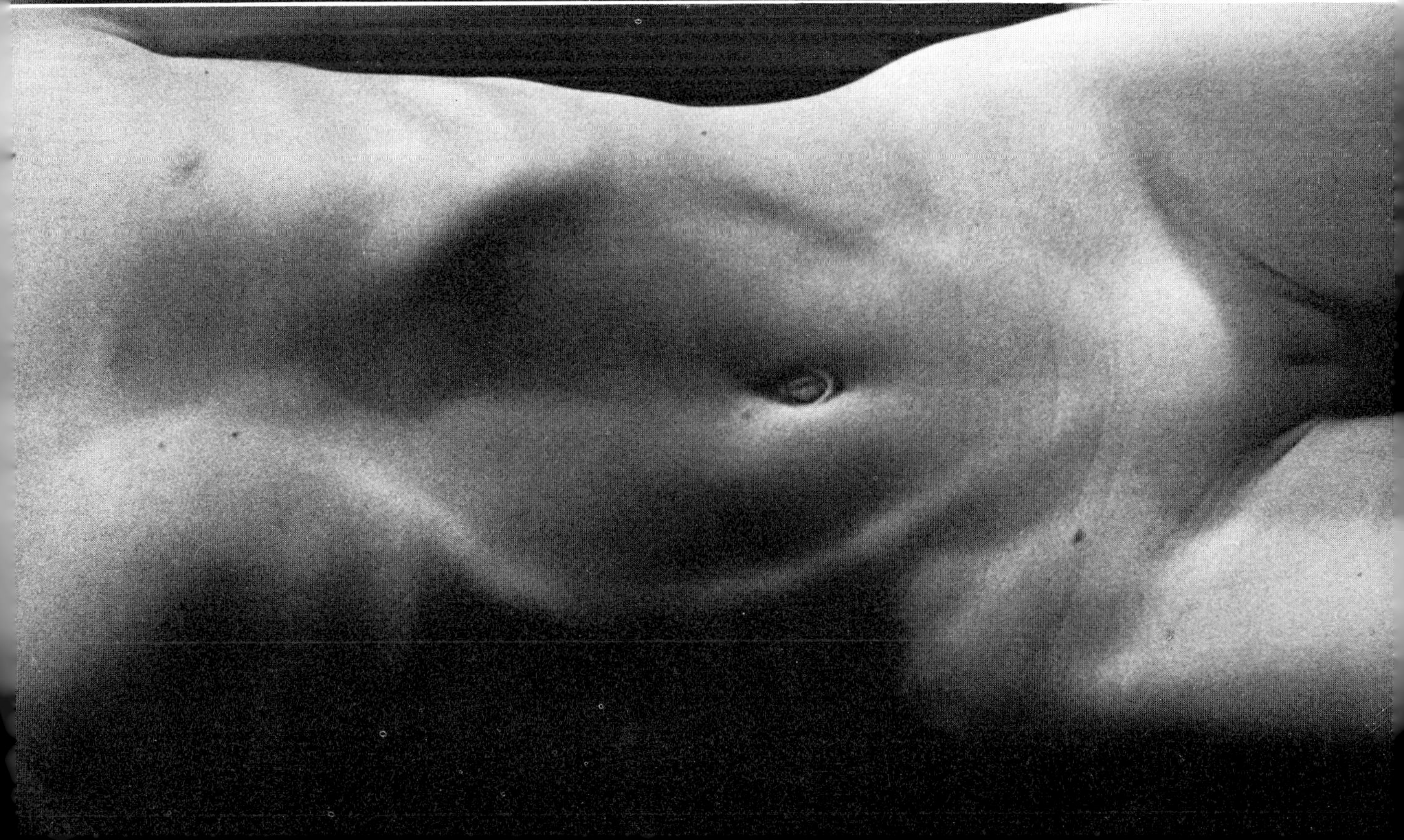

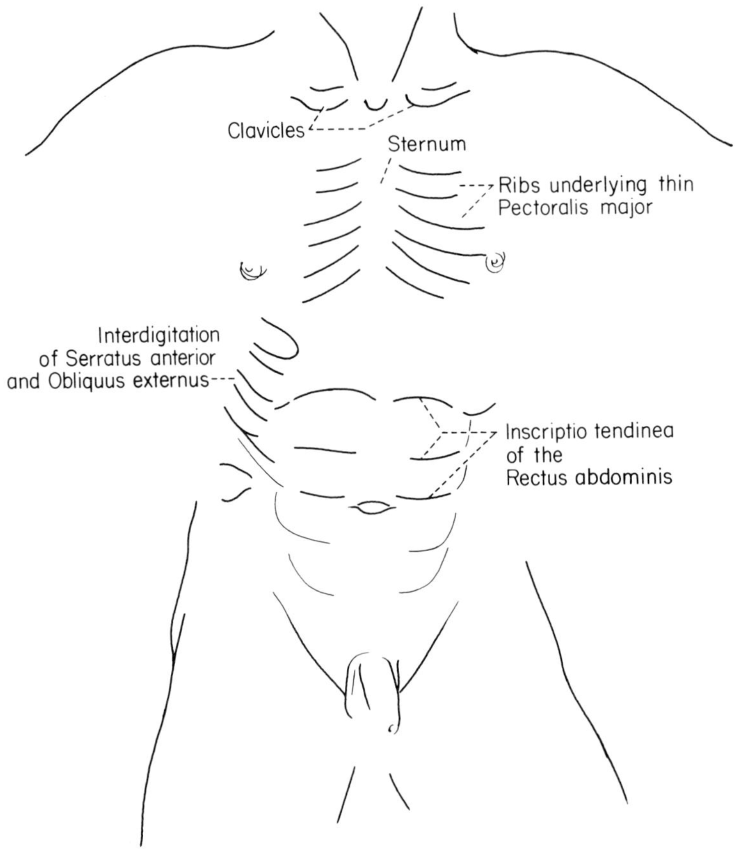

Figure 47 (♂ 7)

The very thin subject shows the "washboard" effect of the abdominal wall where the rectus abdominis is divided (above the umbilicus) into three parts by tendinous inscriptions. (Compare with Fig. 41.) On the chest wall the ribs show through the pectoralis major which is still not well developed.

Figure 48 (♀ 7)

After taking a full breath and closing the glottis it is possible to alternately raise the abdominal wall (contents pushed up by diaphragm) and the ribcage (lifted by the intercostales externi) without breathing.

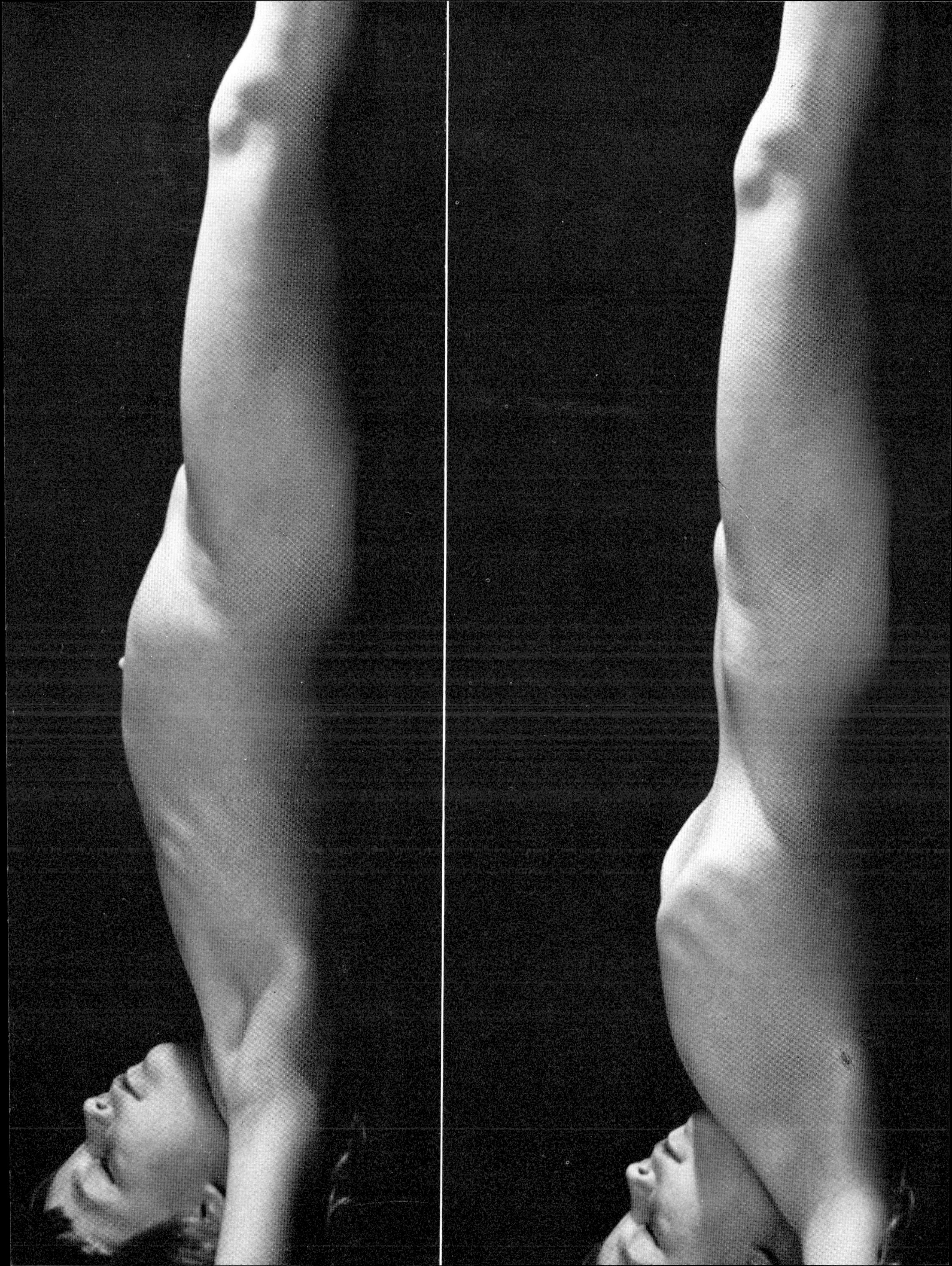

Figure 49 (♀ 17)

By inhaling against a closed glottis the subject is able to flatten the abdominal wall until it actually becomes concave. This girl was able to perform twelve sit-ups with bent knees. The position of the linea alba is marked by a line of downy dark hair.

Figure 50 (♀ 13)

Again the concave abdomen with a well-defined costal margin and pelvis (crista iliaca anterior superior and pubis) in a ballet dancer who was able to perform more than thirty sit-ups with bent knees.

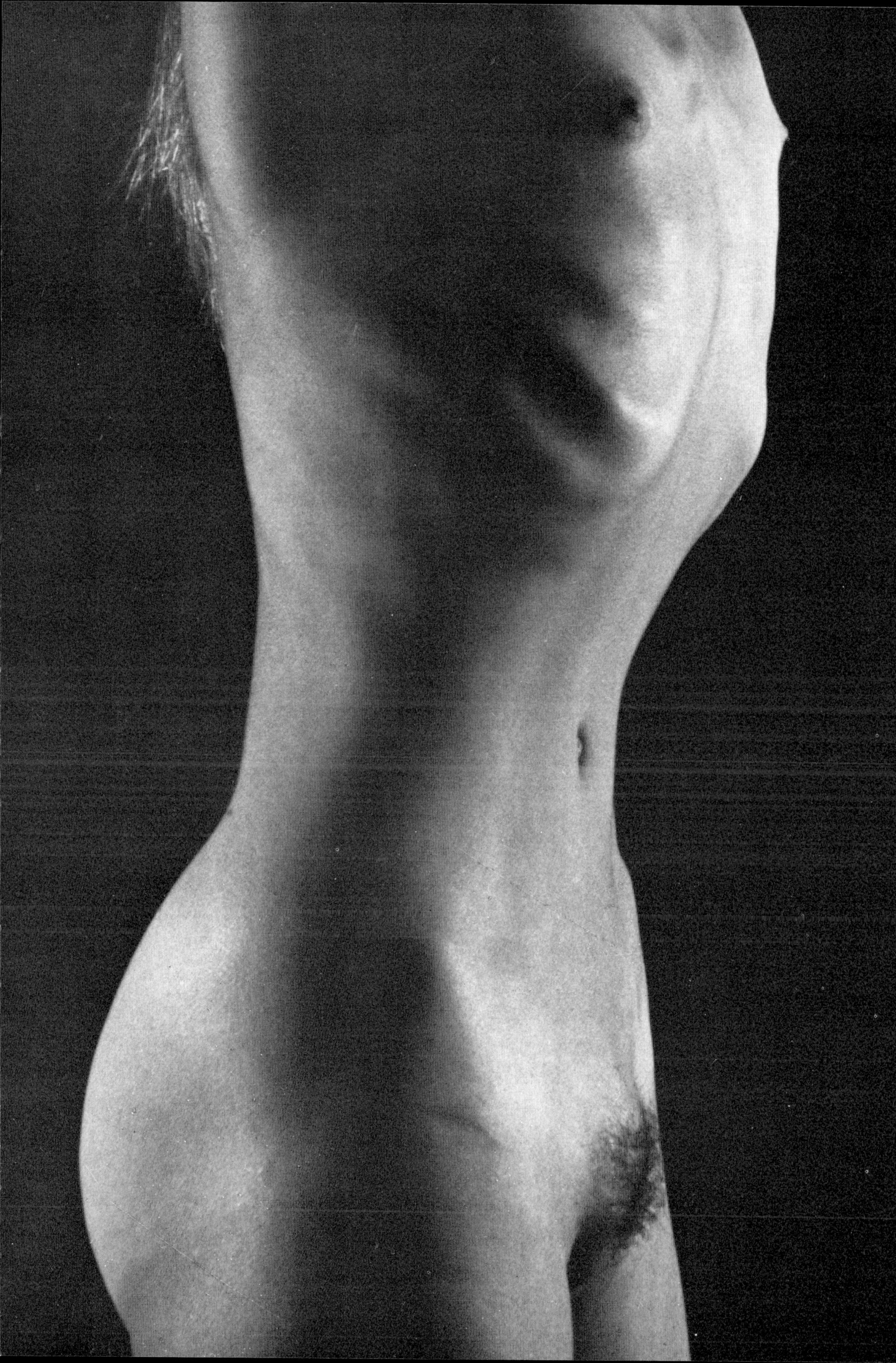

Figure 51 (♂ 72)

Although the subject had comparatively strong abdominal muscles (thirty sit-ups with bent knees) the abdominal contents and adipose tissue prevented him from flattening the abdominal wall (compare with Figs. 49 and 50). The skin has lost some resilience as shown by the clothing marks around the waist which remained for longer than twenty minutes after removal of the clothing.

Figure 52 (♀ 26)

Same subject as depicted in Figure 37 showing post-pregnancy stretch marks on the abdomen. The clothing marks remained visible for longer than twenty minutes. The right breast is lifted by the underlying pectoralis major and by the overlying skin. Axilla is filled up by adipose tissue. It is difficult to discern any bony landmarks.

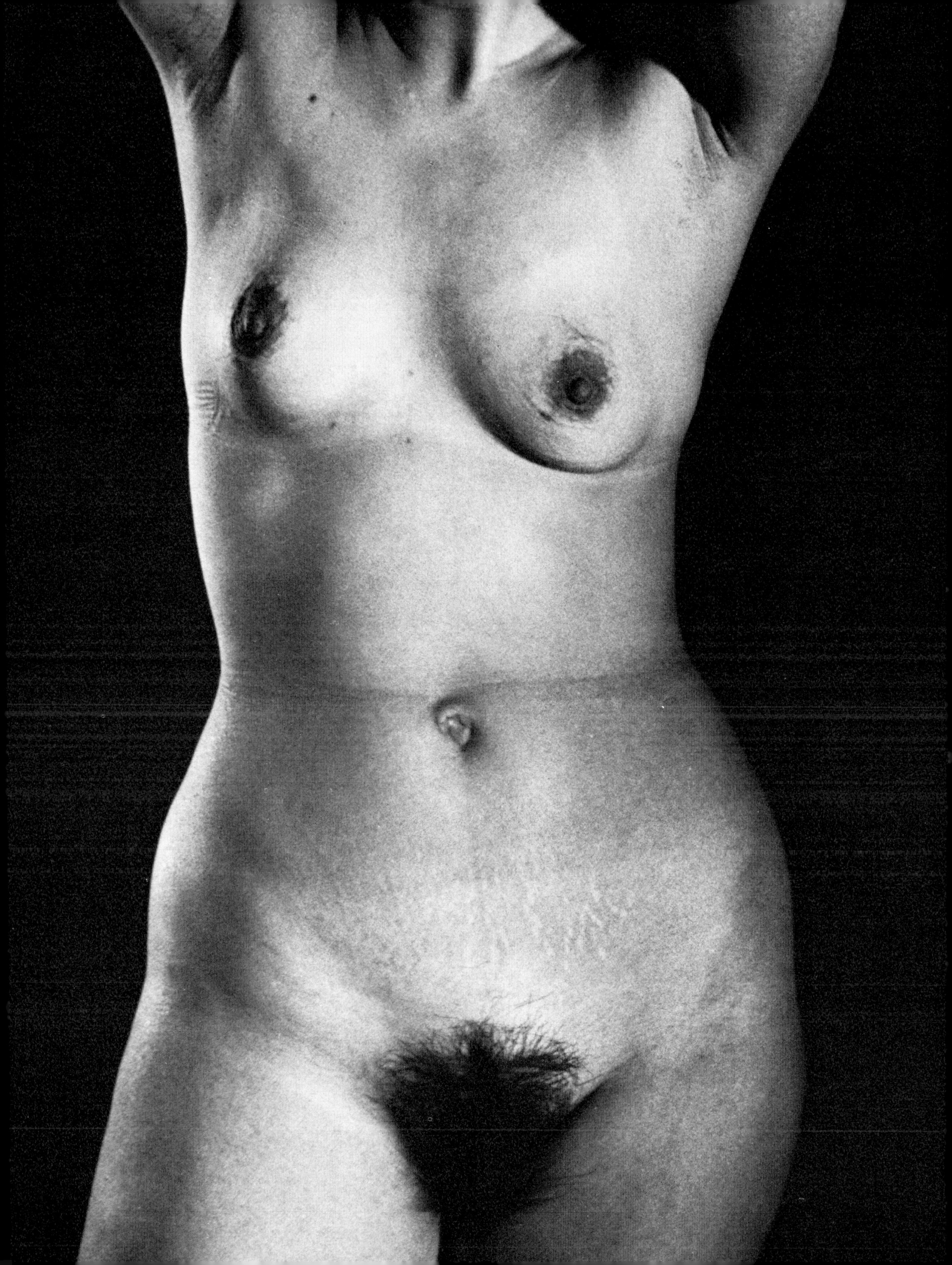

Figure 53 (♂ 9)

The subject exhaled in a half-sitting position, flattening his abdominal wall, and then was able to raise the pencil-thin lower border of the external oblique muscle. This border is free from the underlying tissue and, formerly a thick edge of the external oblique aponeurosis, is called the inguinal ligament.

The author has not seen any other subjects who were capable of voluntarily demonstrating this ligament thus clearly.

Figure 54 (♀ 10)

These identical twins show the venous pattern on the abdomen and thigh (infrared photograph). The girl on the left performed a series of vigorous exercises including sit-ups with bent knees. The girl on the right rested quietly in a cool room. The difference in surface hyperemia was clear, although pre-exercise photographs showed no difference.

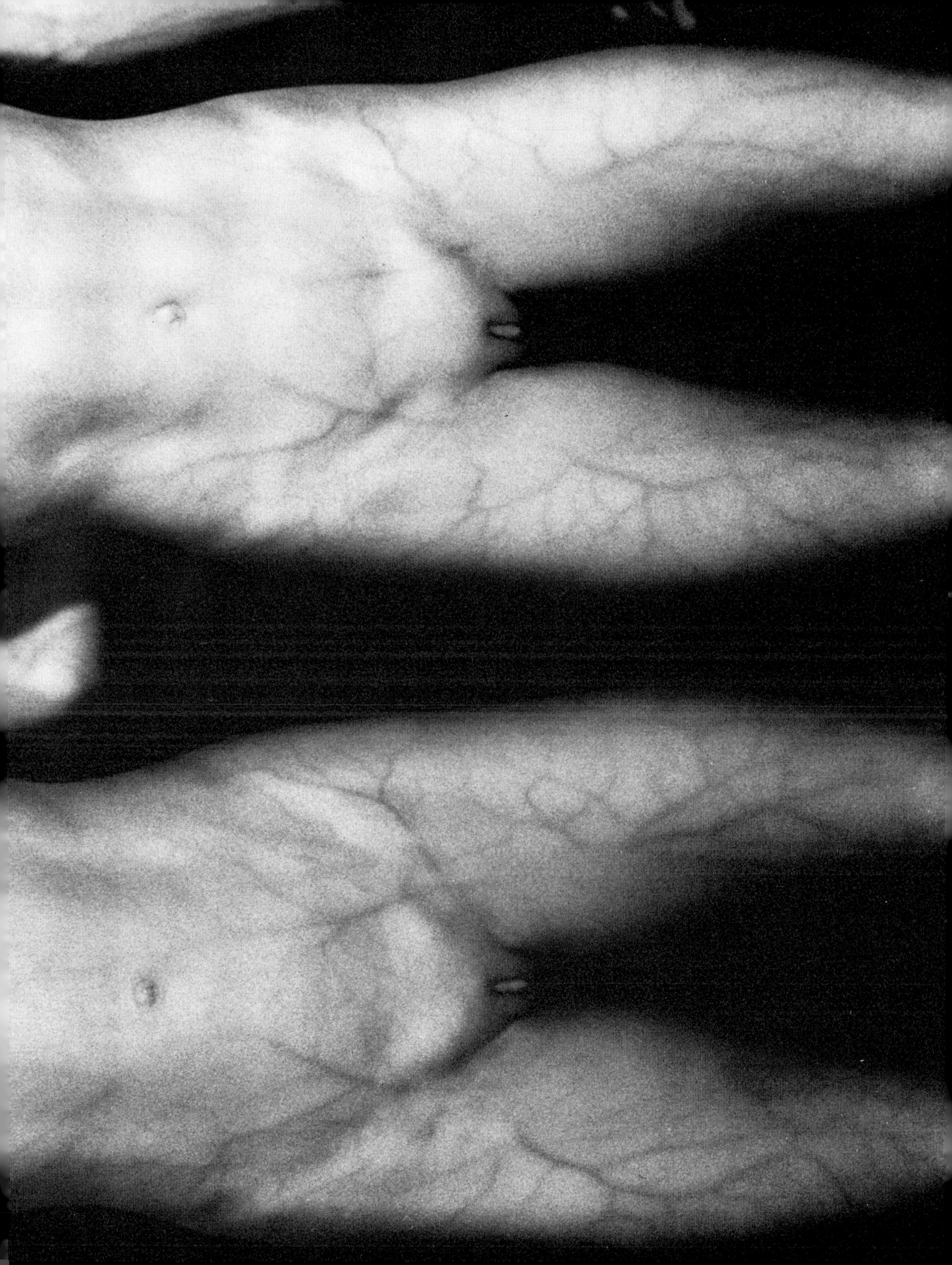

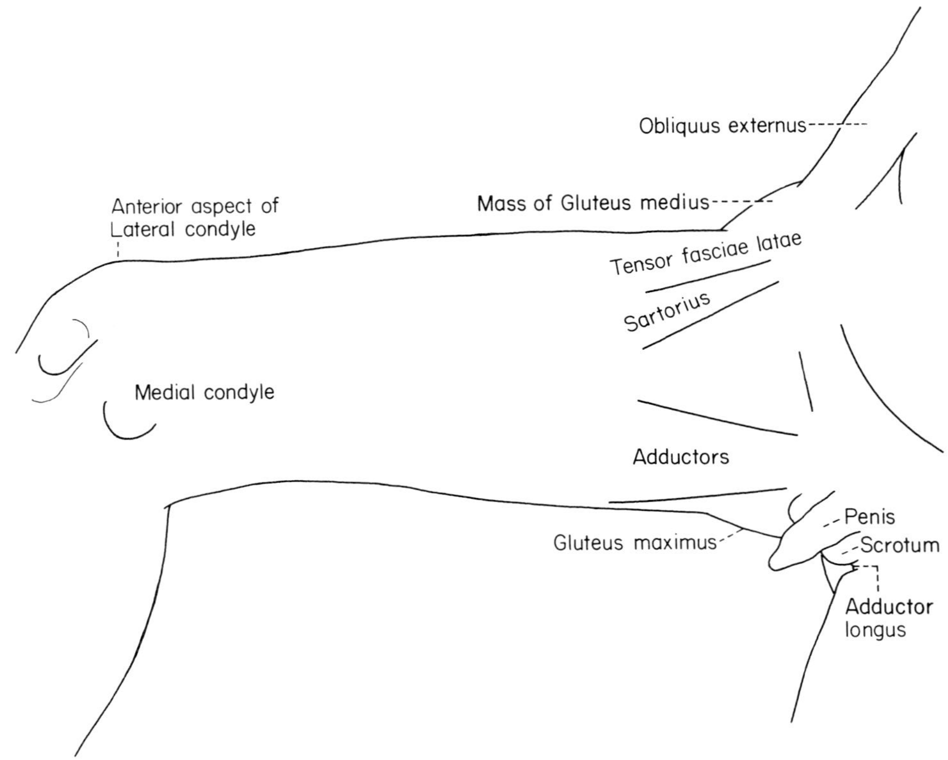

Figure 55 (♂ 7)

The triangle formed by the inguinal groove (groin), the medial border of the sartorius and the bulge of the adductor longus is clearly seen when the femur is lifted against resistance while in an outward rotated position. The floor of this femoral triangle is formed by the pectineus and the iliopsoas which, in thin subjects, can be palpated.

The testicles have not yet descended.

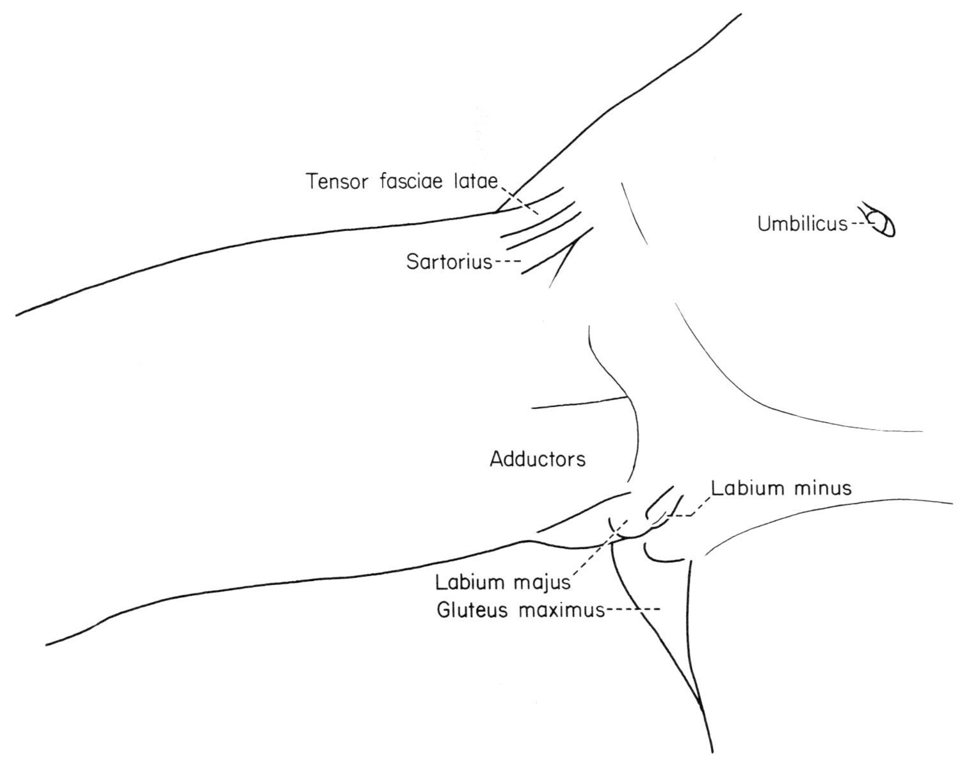

Figure 56 (♀ **11, late maturing)**

The inguinal groove ends medially behind the mass of the adductor group at a place where the gluteal fold also ends medially. At the spina iliaca anterior superior the tensor fasciae latae and the sartorius are seen to originate. These two muscles split up to reveal the rectus femoris at approximately three inches below their origin. Here the tensor runs lateral-ward and the sartorius medial-ward on the thigh.

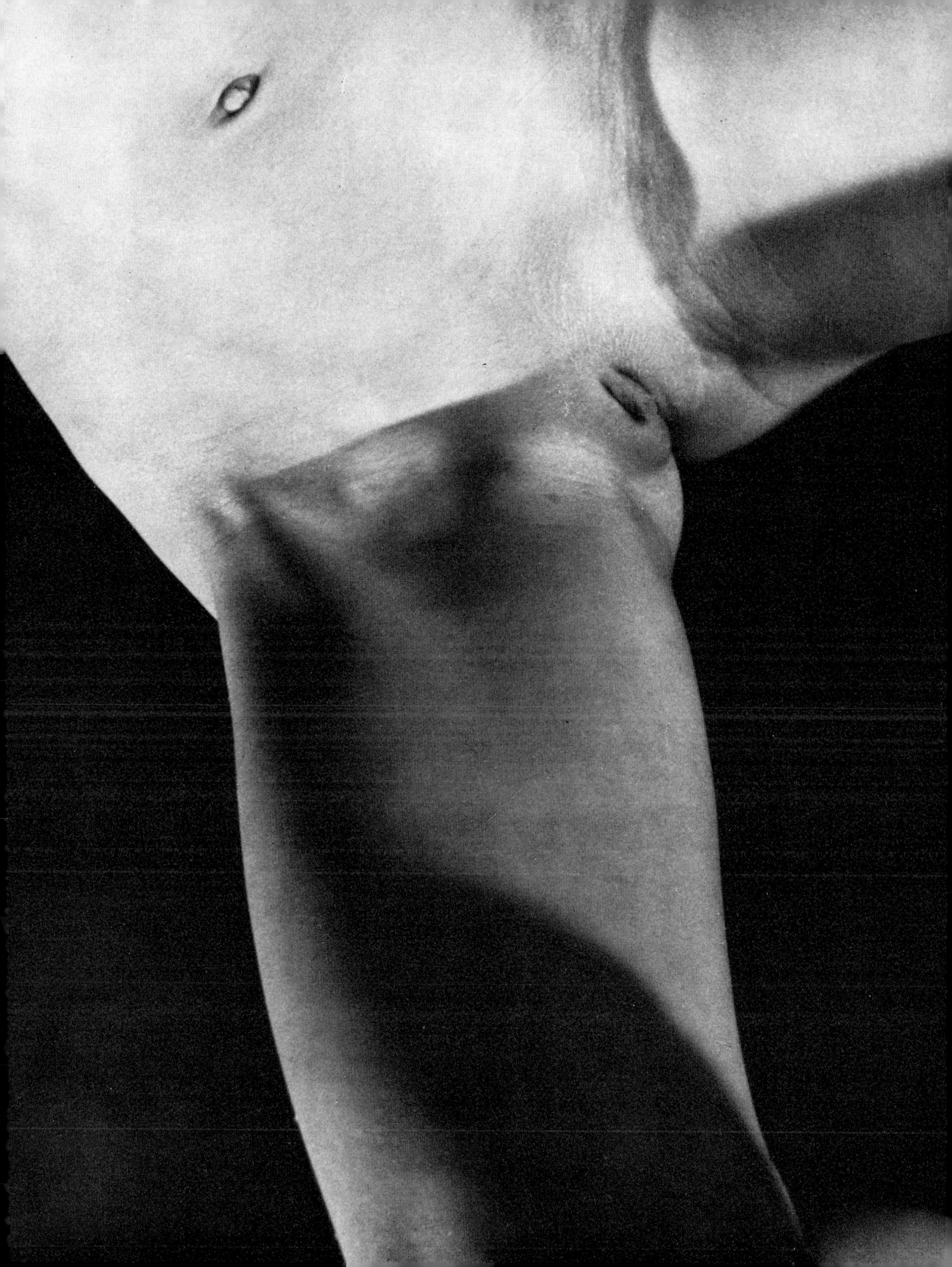

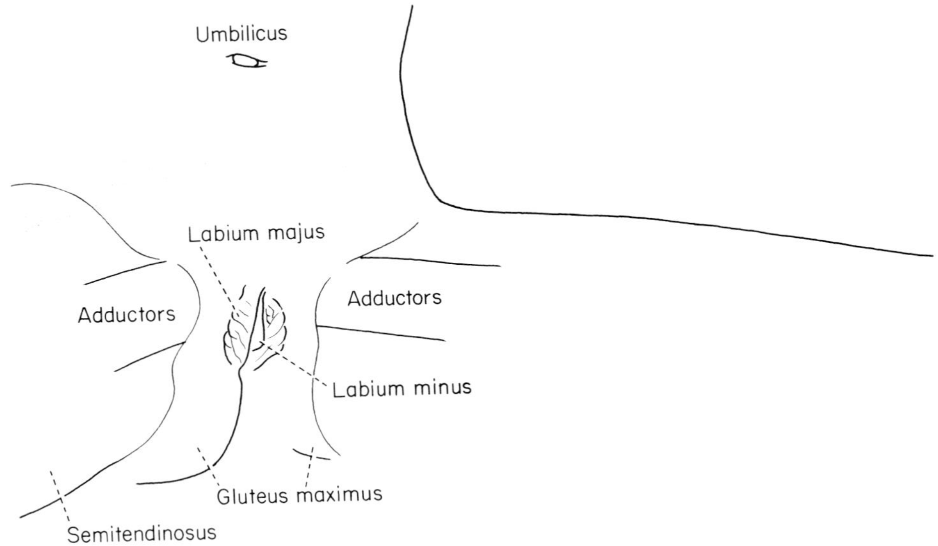

Figure 57 (♀ 8)

The adductor longus (and the gracilis) originates on the os pubis (pubic bone) which is mostly covered with a fat pad. The semitendinosus originates on the os ischii (sitting bone) which is covered by the buttocks. These muscles can be seen in thin subjects when they attempt to bring the thighs together against resistance.

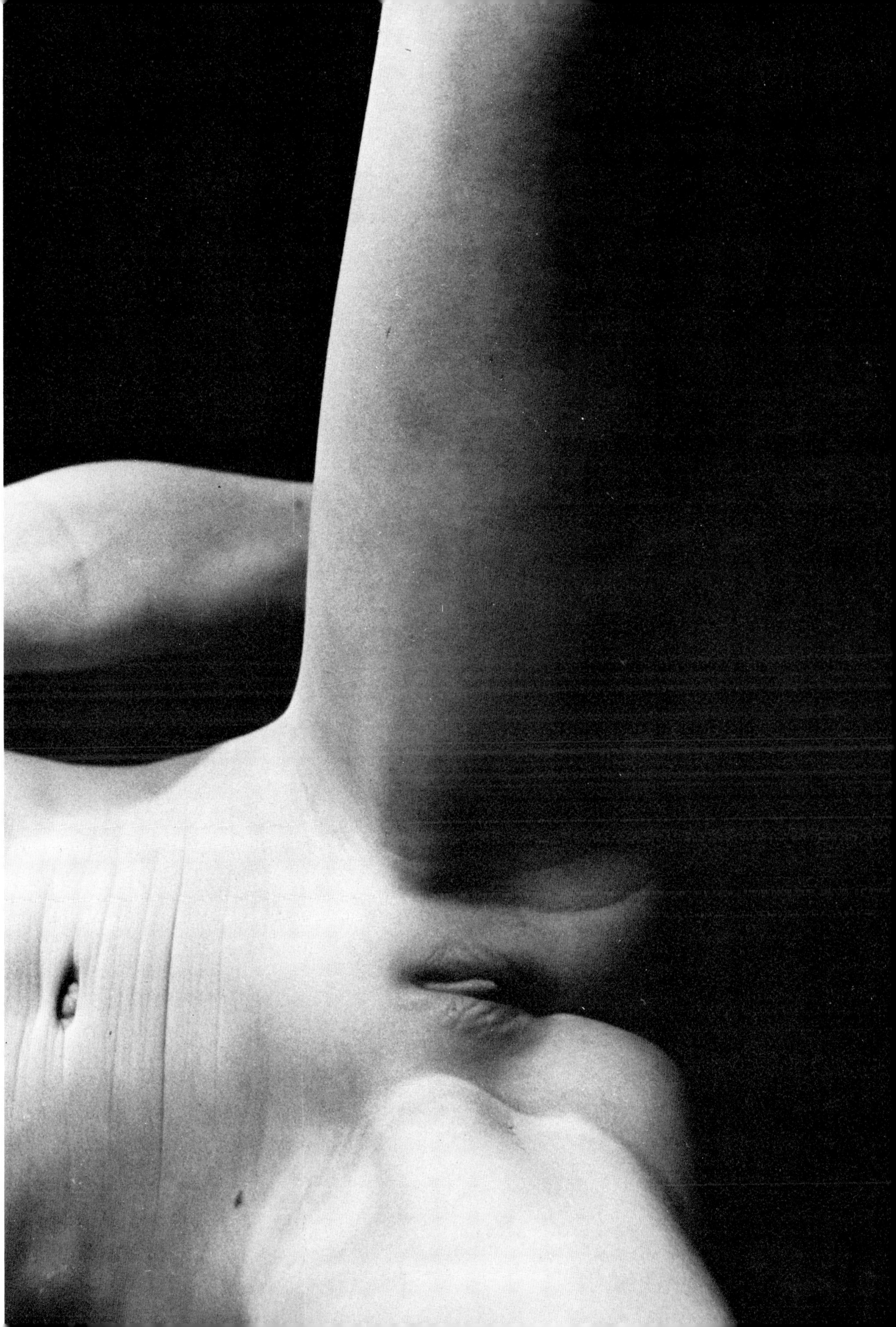

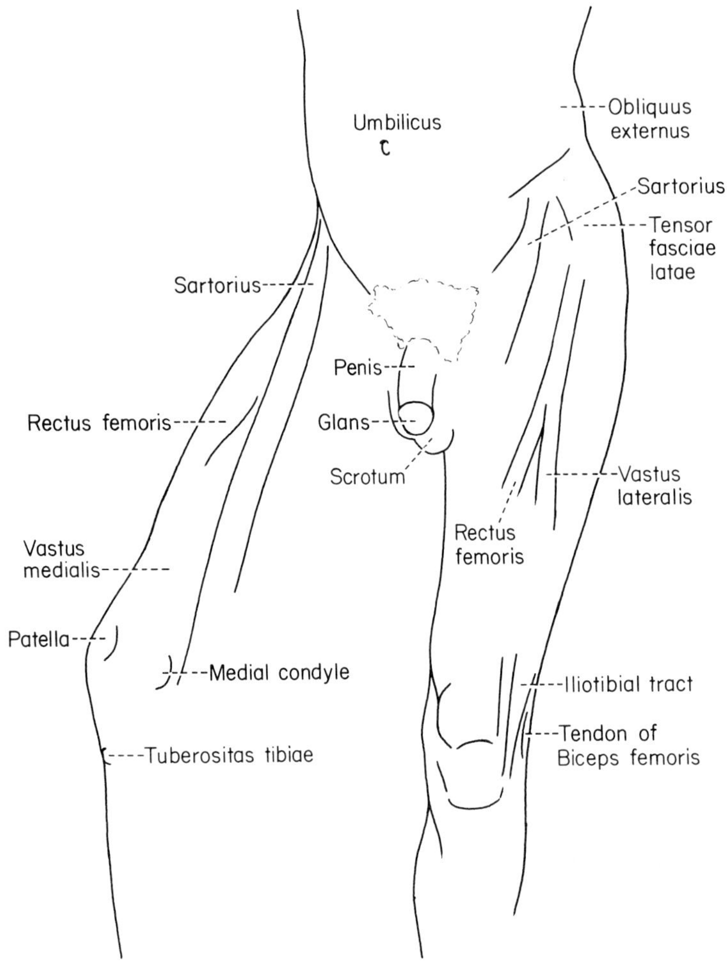

Figure 58 (♂ 26)

This thin subject demonstrated the sartorius by attempting to push the right foot forward against the friction of the floor while the leg was in an outward rotated position.

The male pelvis is generally less wide than the female pelvis. This results in a steeper V-form of the inguinal groove (see also Figs. 41 and 52).

The outstanding vein on the medial side of the thigh is the saphena magna.

Supporting the body weight on the left leg tilts the pelvis to the right; the balance is retained by the compensatory shift of the shoulder region to the left (see also Fig. 59).

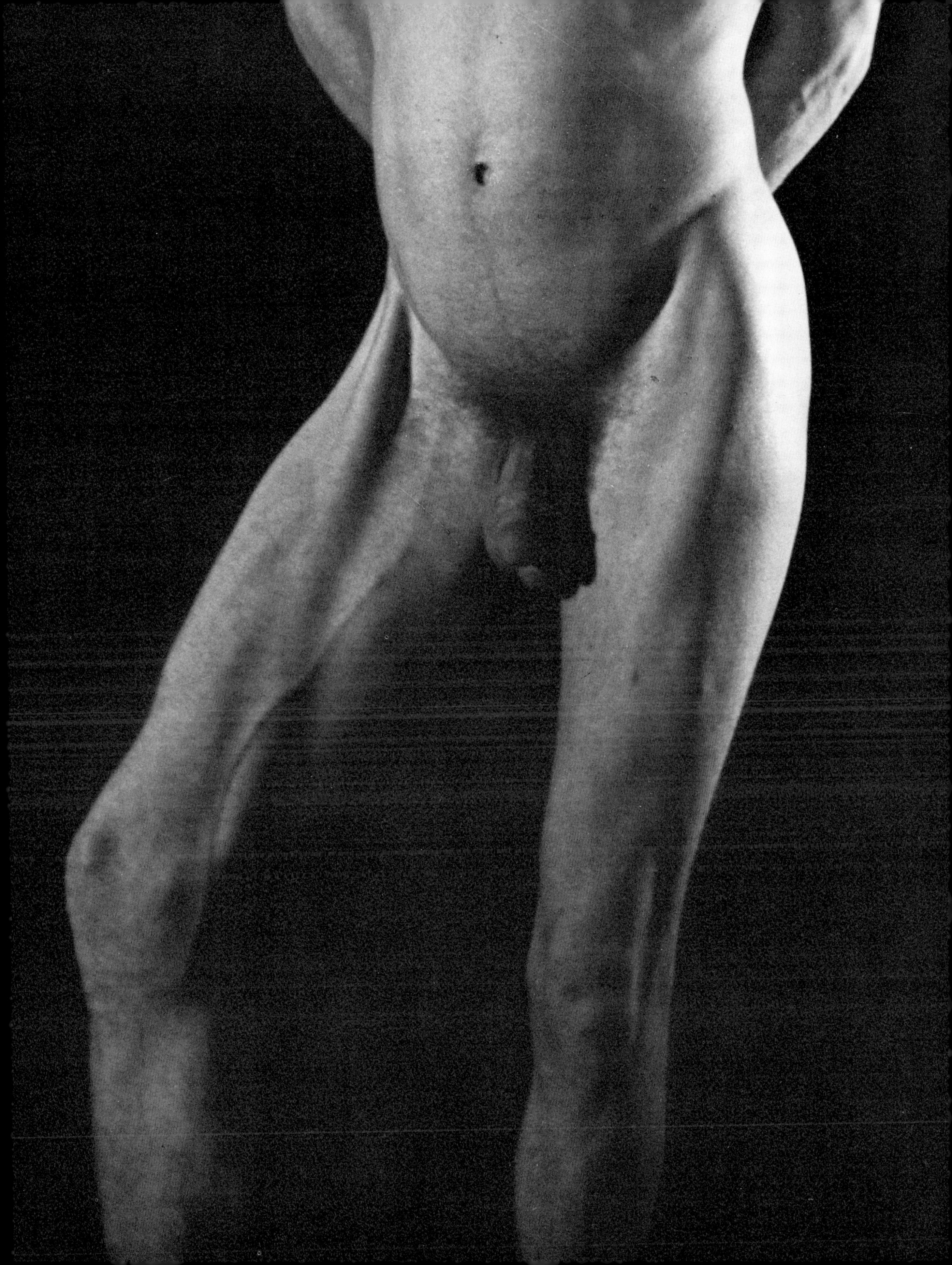

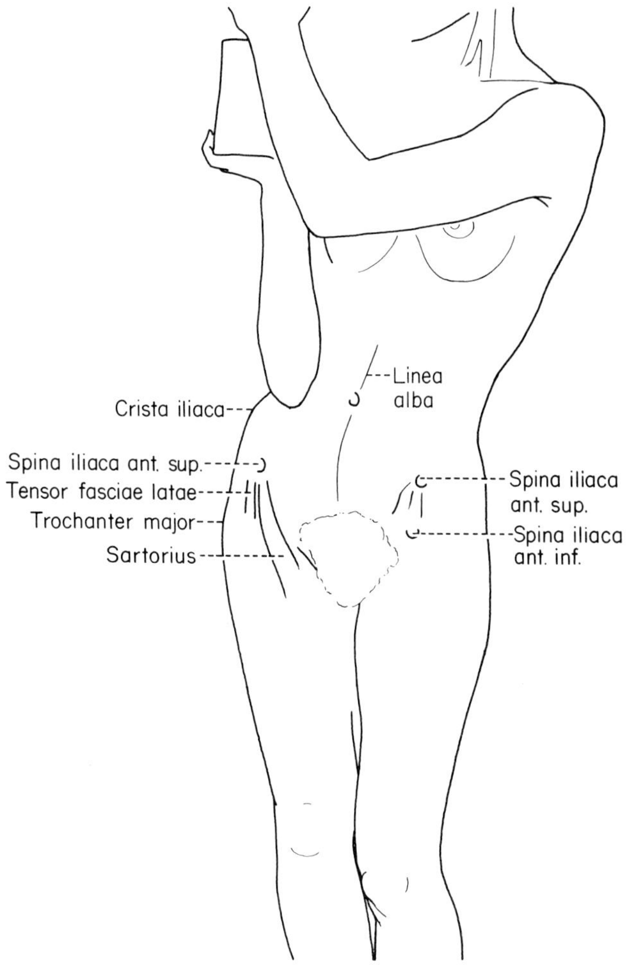

Figure 59 (♀ 17)

The body weight and an additional ten-pound weight are supported on the right leg. The resulting lateral tilt of the pelvis to the left side must be compensated for by curving the spine to the right in order to maintain equilibrium with the center of gravity over the area of support. The right elbow is supported by the iliac crest thus minimizing the effort of the right shoulder muscles in holding the ten-pound weight.

In this thin subject the spina iliaca anterior inferior is visible as a slight bulge underneath the sartorius muscle. Downy dark hair grows along the linea alba up to the umbilicus.

Figure 60 (♀ 17)

In the present subject (slightly on the chubby side) the flattening of the abdominal wall brings out the spina iliaca anterior superior. The inguinal groove is erased but the accompanying skin creases are visible. Fat deposits on the gluteus medius and the adductors exaggerate the size of these muscles.

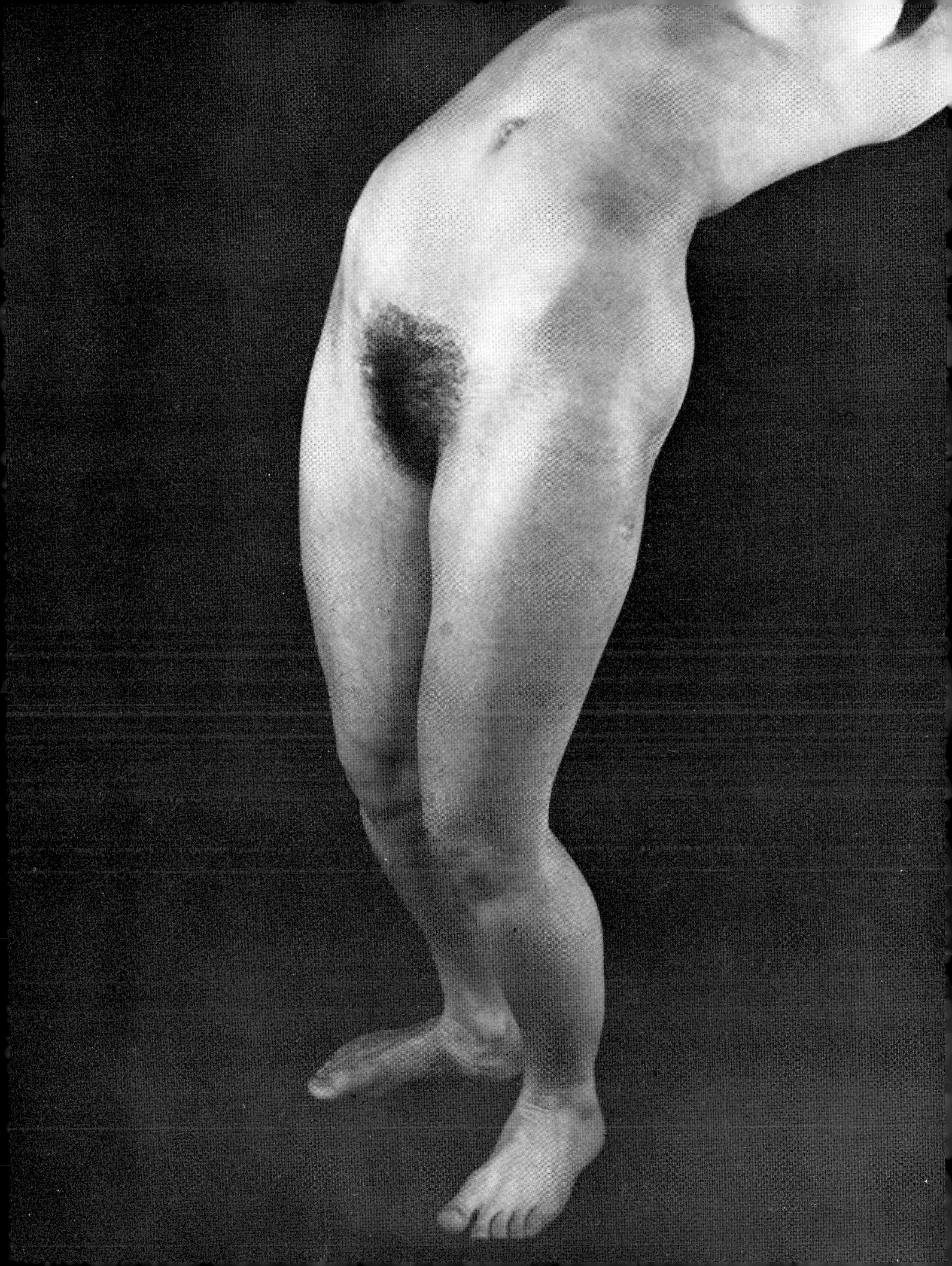

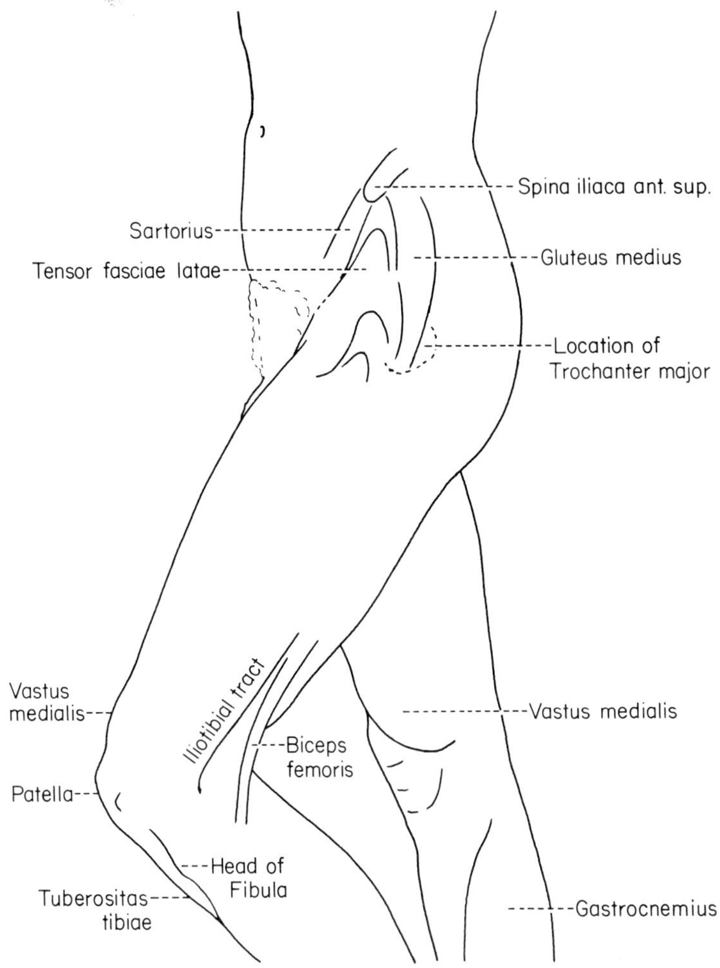

Figure 61 (♂ 23, tall, thin)

The subject lifted a heavy metal boot with the left foot while holding on to an overhead beam for balance. The large veins on the abdominal wall are branches of the epigastrica superficialis.

Figure 62 (♀ 11)

Hip abductors are still visible on this young, thin subject. Fat deposits will disguise these muscles later (see also Figs. 31 and 32) and also the angle made by the upper border of the gluteus maximus on the lower back contour. The bulge on the os pubis is due to pre-pubic fat deposits (the so-called mons pubis, see also Fig. 57).

Figure 63 (♀ **11**)

The same subject as shown in Figure 62 in a one-leg stand, with extension and abduction in the left hip joint. Since extension is limited the pelvis must be tilted forward on the right hip joint to raise the left leg backward to a horizontal position (as in figure-skating). The trochanter major and crista iliaca anterior superior are clearly visible in this thin subject.

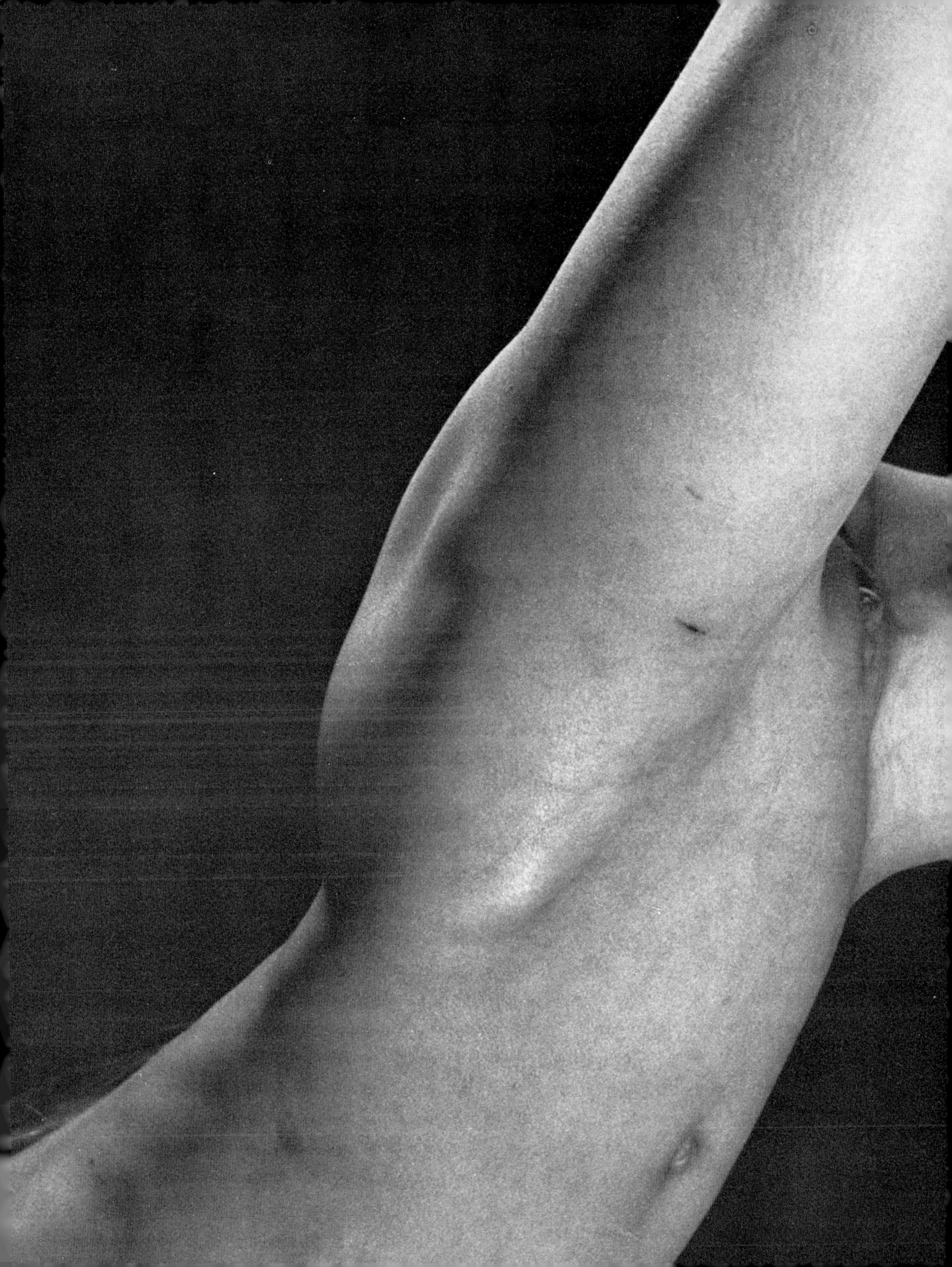

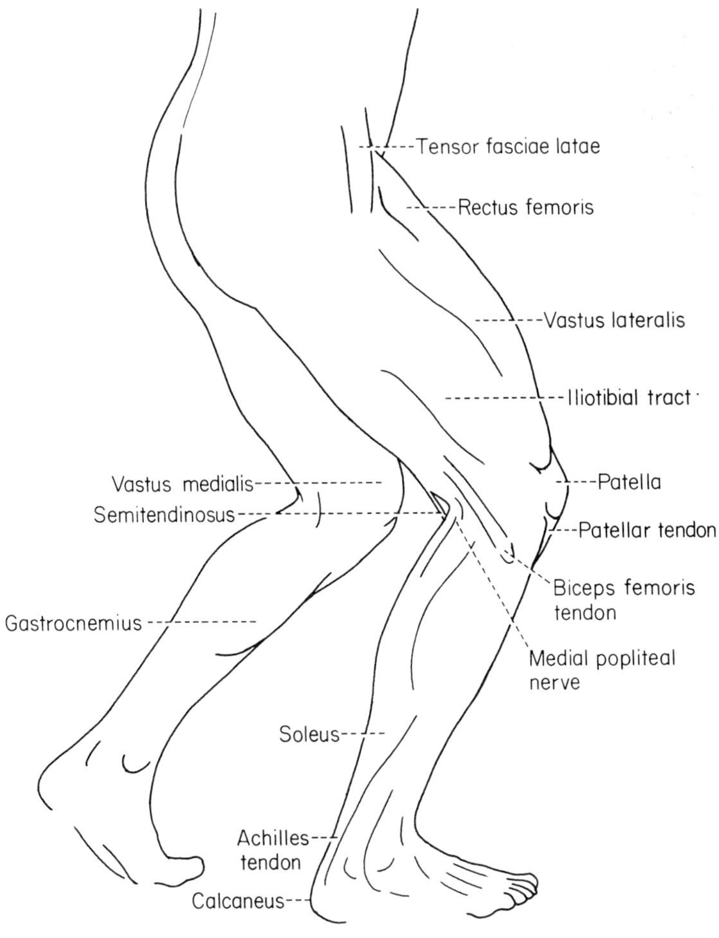

Figure 64 (♂ **23)**

Years of training and competition have developed the leg muscles of this bicycle racer. Although the circumference of the thigh was much greater than that of the average male of the same age, the subject had little subcutaneous fat.

The hamstrings (semitendinosus, semimembranosus, and biceps femoris) are contracted to provide extension in the hip joint, while the quadriceps (rectus femoris, vastus lateralis, intermedius and medialis) extend the knee joint. The hamstrings also flex the knee, while the rectus femoris also flexes the hip joint. This contrary action is neutralized because of certain mechanical advantages in the total lever system so that both muscle groups can contract to act against gravity (Lombard's paradox).

Figure 65 (♂ **26**)

Extreme flexion in the hip joint is possible when the hamstrings are given some slack by flexing the knee. The muscle bulk of the tensor fasciae latae is clearly visible.

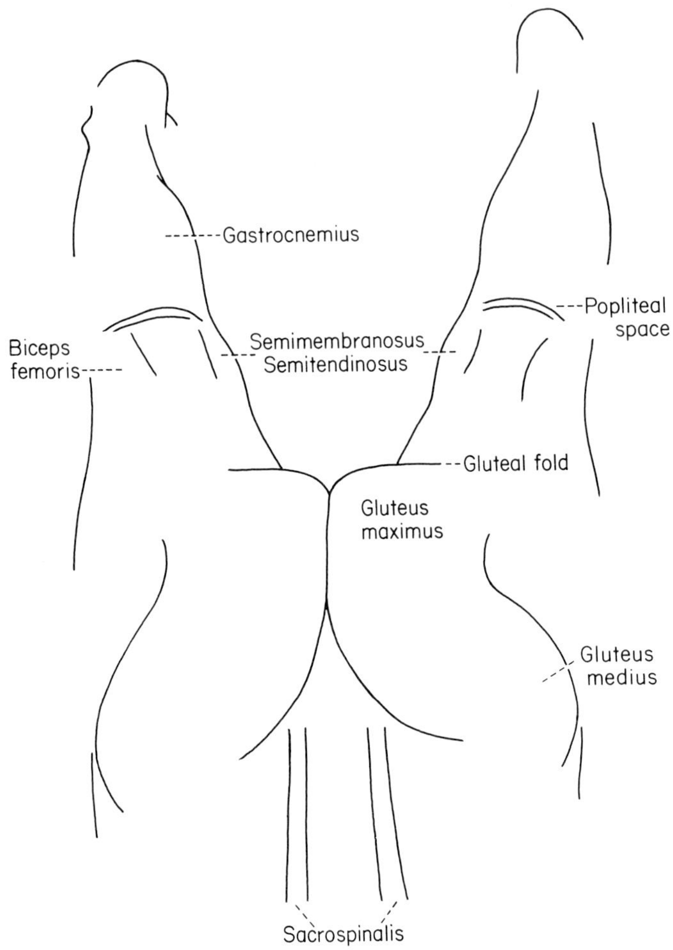

Figure 66 (♂ 8)

Lying face down the subject extended the hips (lifted the legs off the table) and attempted abduction against resistance. The gluteus maximus muscles are contracted vigorously (buttocks pulled together) as well as the hamstrings. The gluteus medius muscles are readily seen in their relation to the lower back since the shadow line approximates their upper contours.

To prevent tilting (backward) of the pelvis because of the levering action of the lifted legs, the sacrospinalis muscles are contracted.

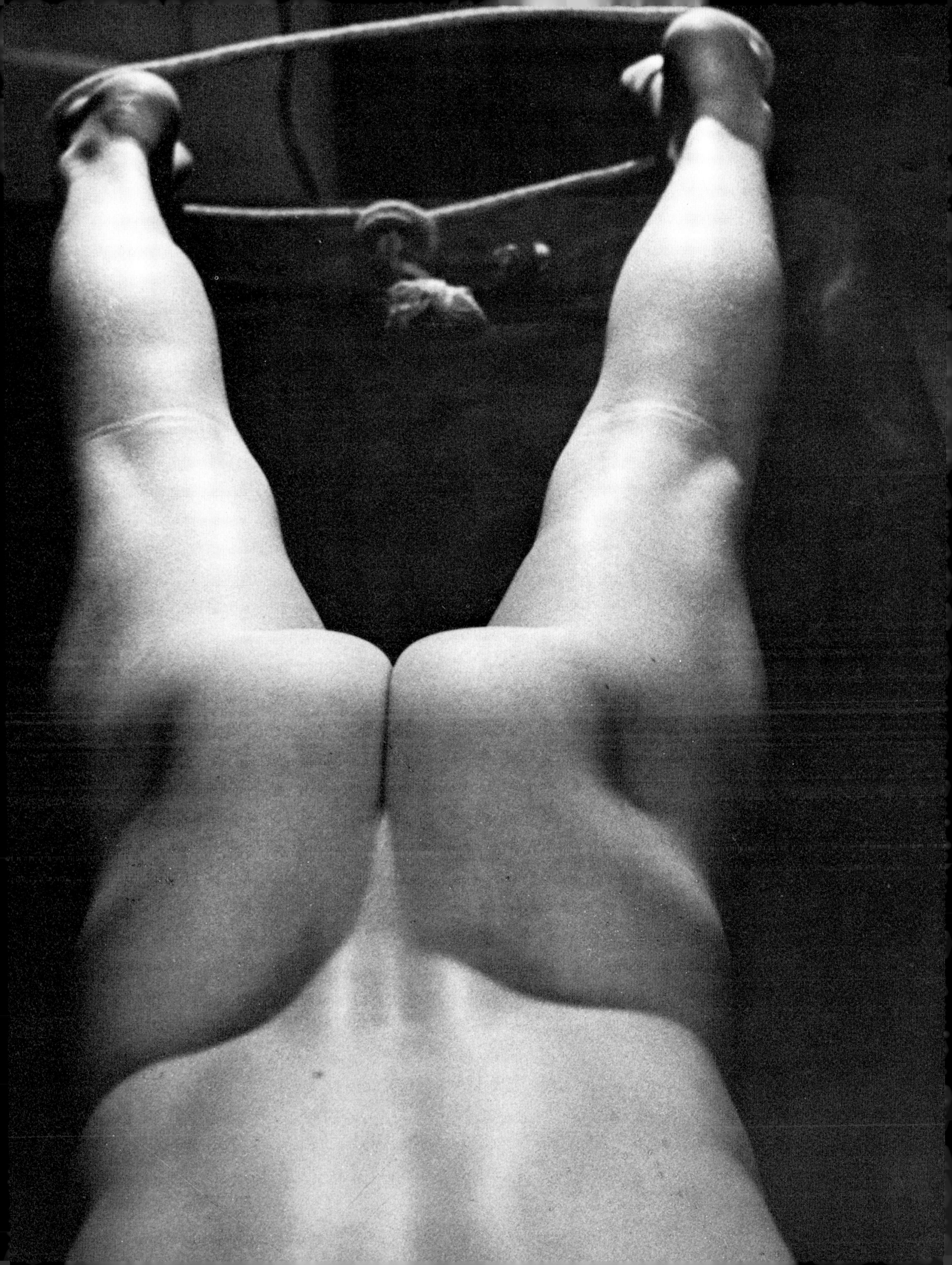

Figure 67 (♀ 26)

The subject raises the legs off the table (as in Fig. 66). Again it is clear that the gluteus maximus is strongly contracted. Other muscle contours are disguised by fat deposits on this mature female.

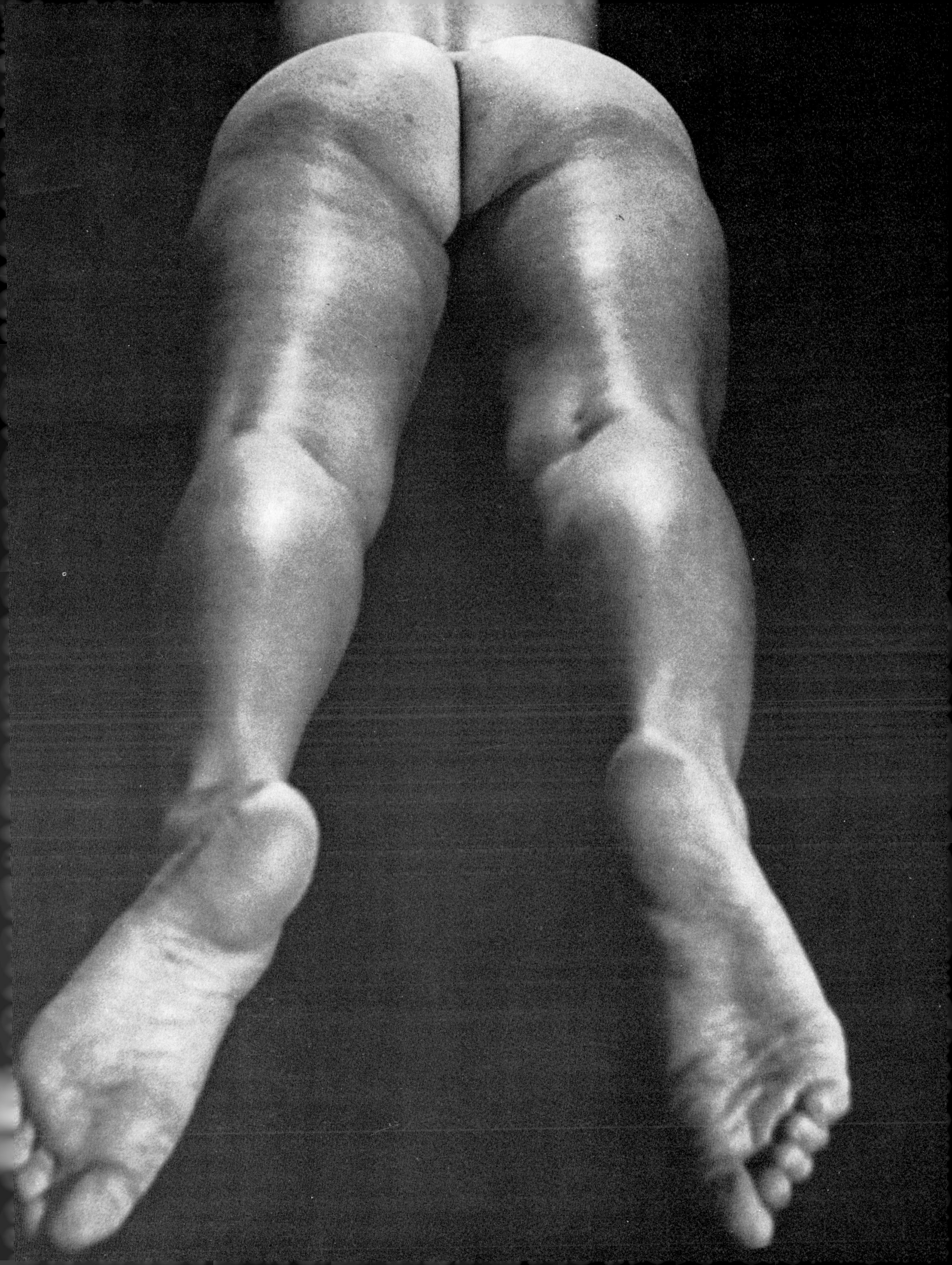

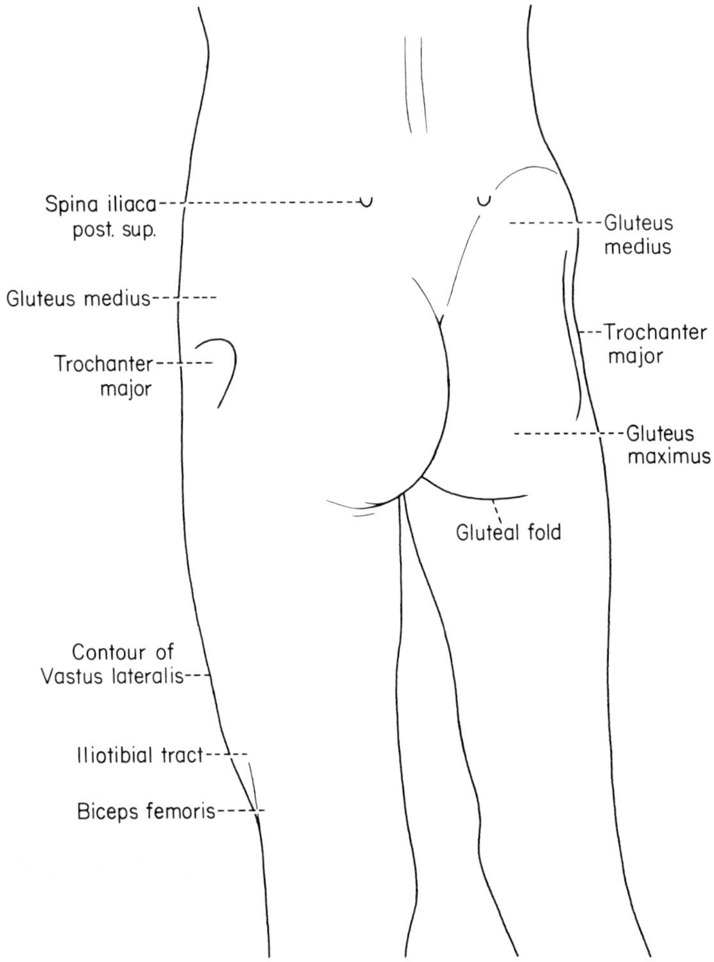

Figure 68 (♀ 12)

A triangle can be seen between the upper end of the gluteal cleft (located at approximately the third part of the sacrum) and the dimples formed by the spina iliaca posterior superior. Asymmetrical tilting in this triangle while standing or walking may be indicative of weaknesses in the gluteus muscles.

In this thin, immature subject the trochanter major can still be observed. The contour of the vastus lateralis (partly covered by the iliotibial tract) is visible.

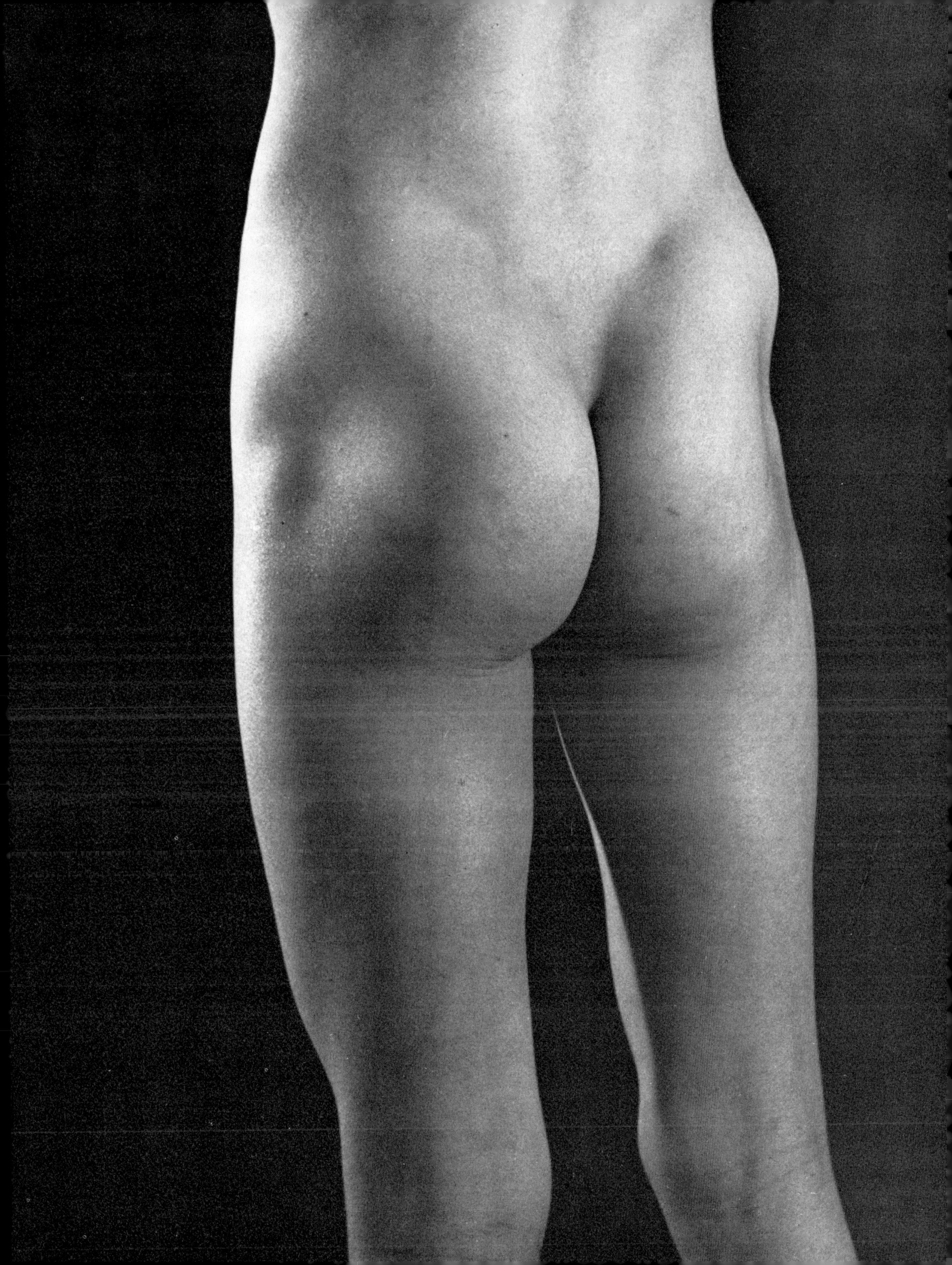

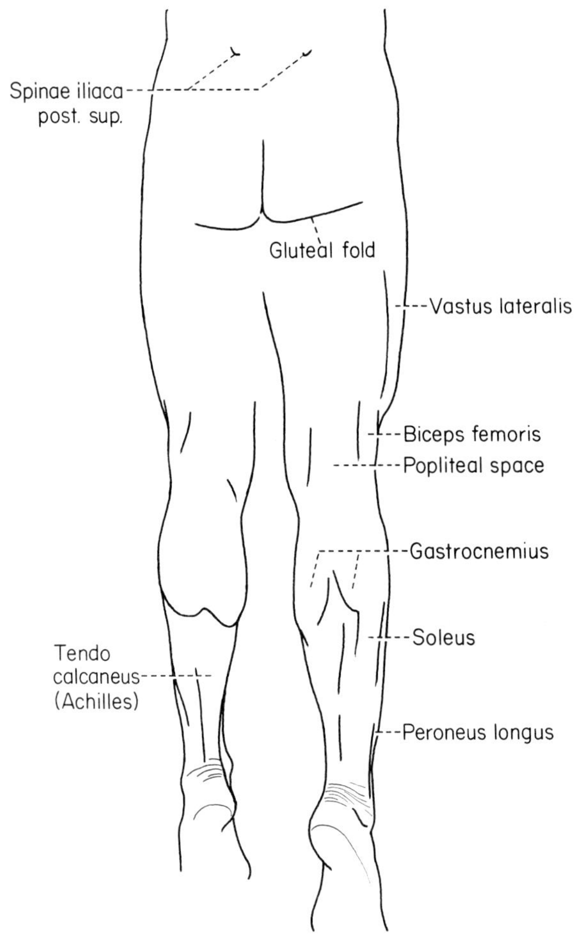

Figure 69 (♂ 23)

The same bicycle racer depicted in Figure 64 shows the highly developed vastus lateralis and gastrocnemius (two muscle bellies). Raising on the toes (plantar flexion in the ankle and foot against resistance) shows the curve (laterally) in the foot sole. The hamstrings are large and do not show their separate contours.

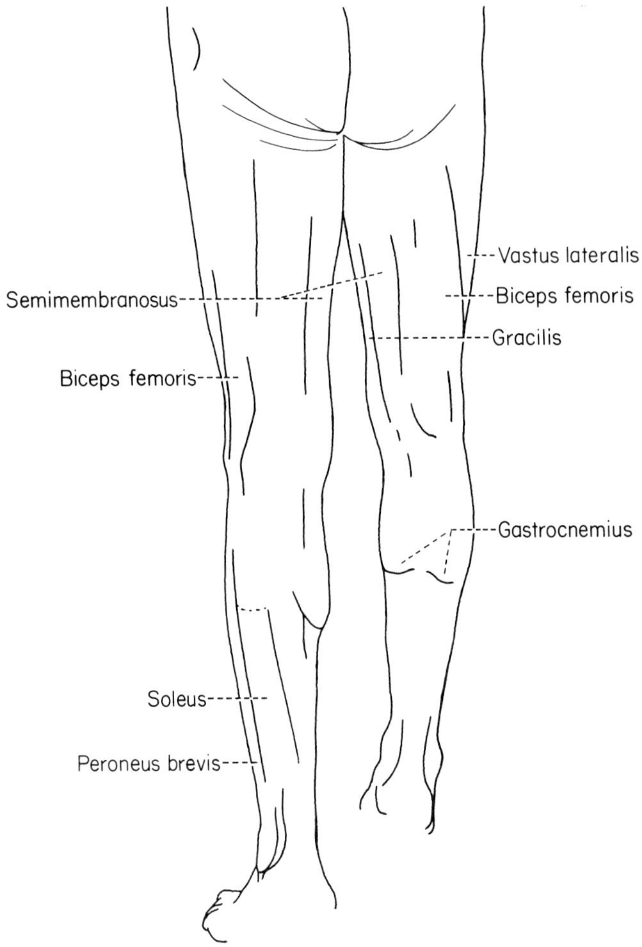

Figure 70 (♂ 72)

This subject had participated in conditioning exercise classes for over fifty years (three times per week for one hour). He was able to run a mile a day at approximately eight miles per hour without undue fatigue. Despite his age he shows definite muscle contours. The buttocks have slightly atrophied. The skin has lost resilience (gluteal fold and clothing marks) and the lower legs show varicose veins.

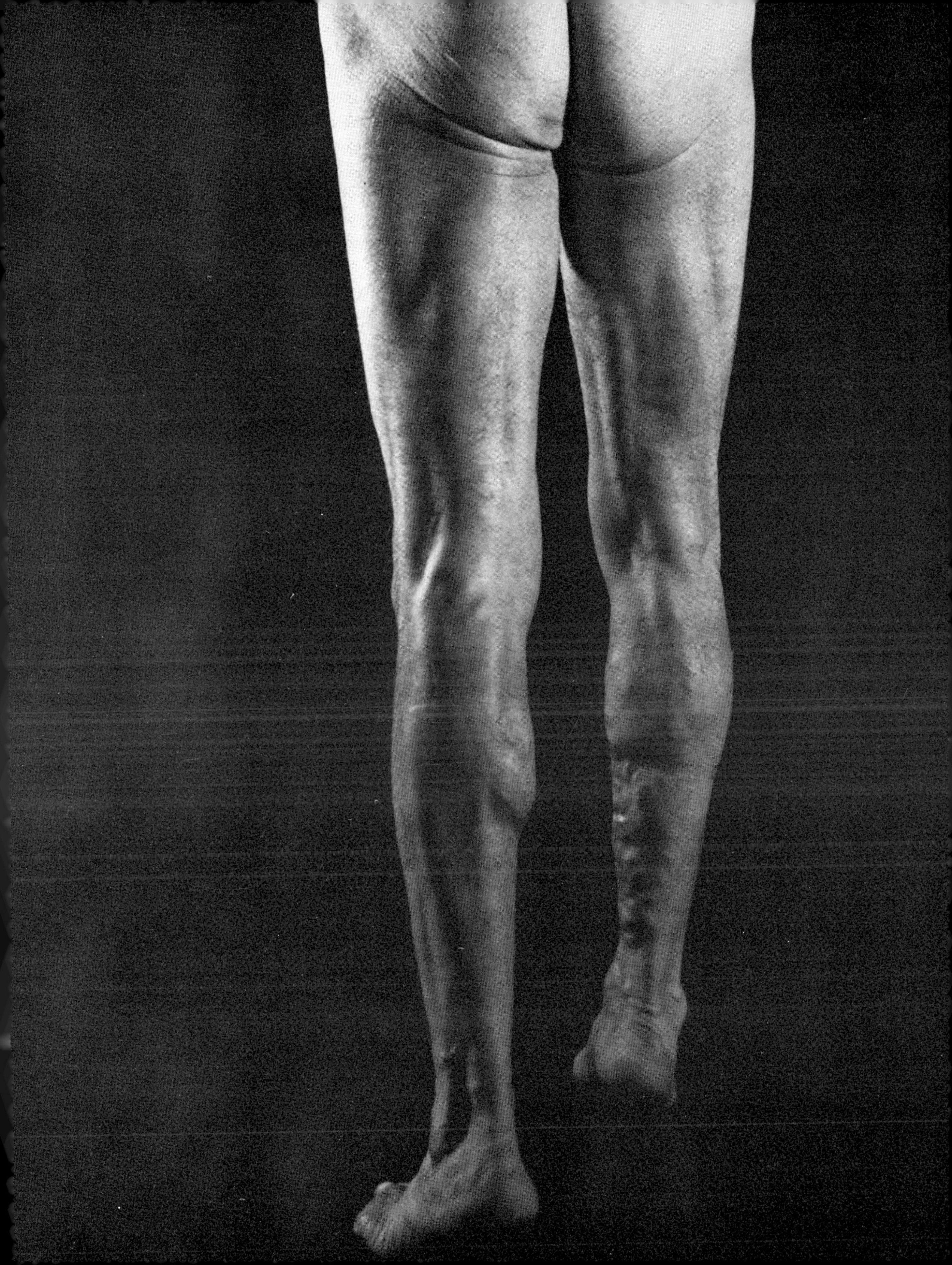

Figure 71 (♀ 44 and 52)

Fat deposits have obscured the underlying muscle contours in the subject on the left (forty-four years old), although she had participated for thirty-two years in calisthenics clubs (three times per week for one hour). The fat deposits over the posterior iliac crest give adult females the long-waisted look as compared to adult males. The trochanter major can only be found by careful palpation, preferably when the hip joint is alternately flexed and extended (compare to Figs. 68, 69).

The subject on the right (fifty-two years old) participated in vigorous calisthenics (three times per week for one hour) regularly for forty-one years. The gluteus maximus has started to atrophy, but the other leg muscles were firm although obscured by fat deposits.

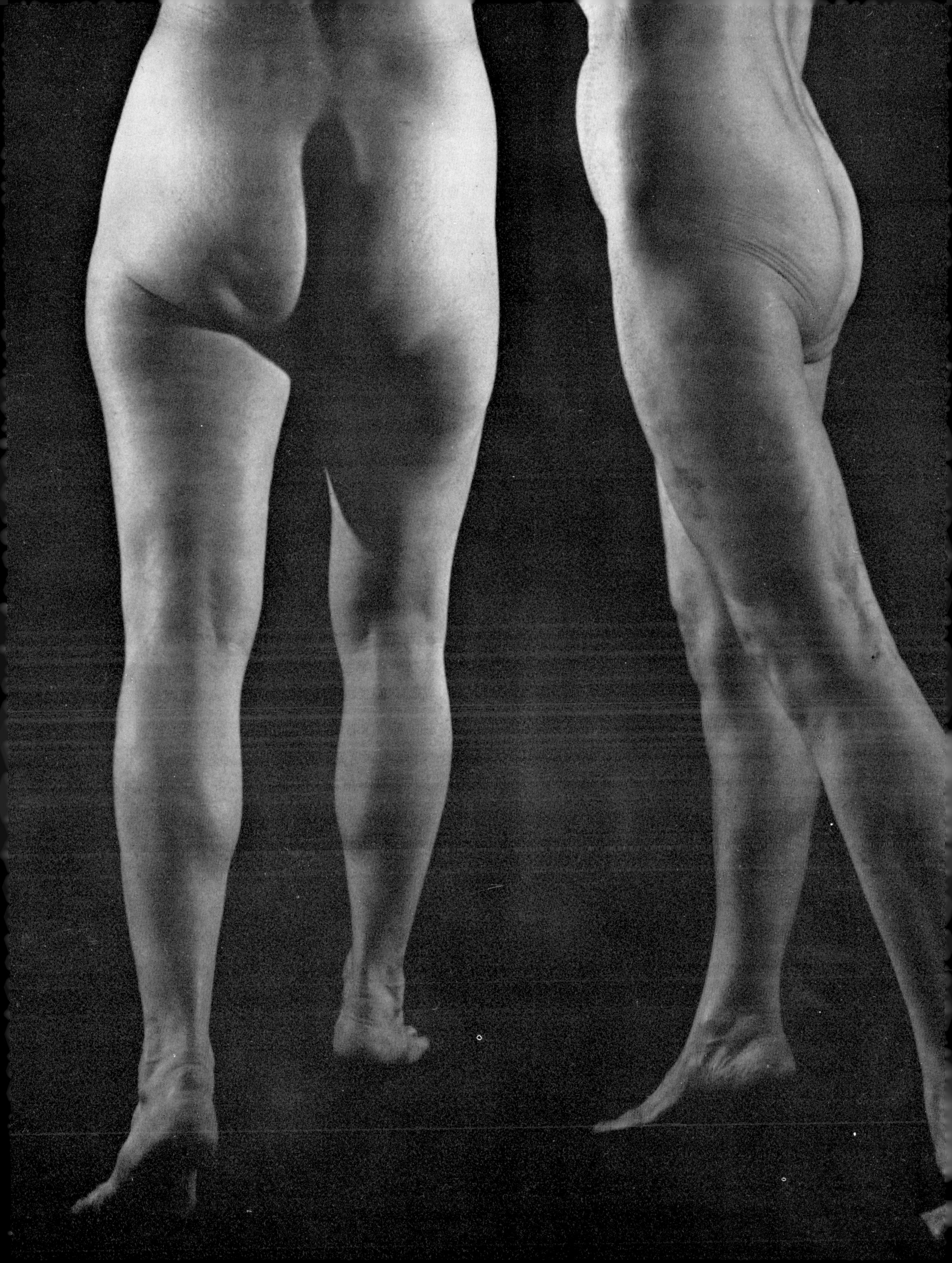

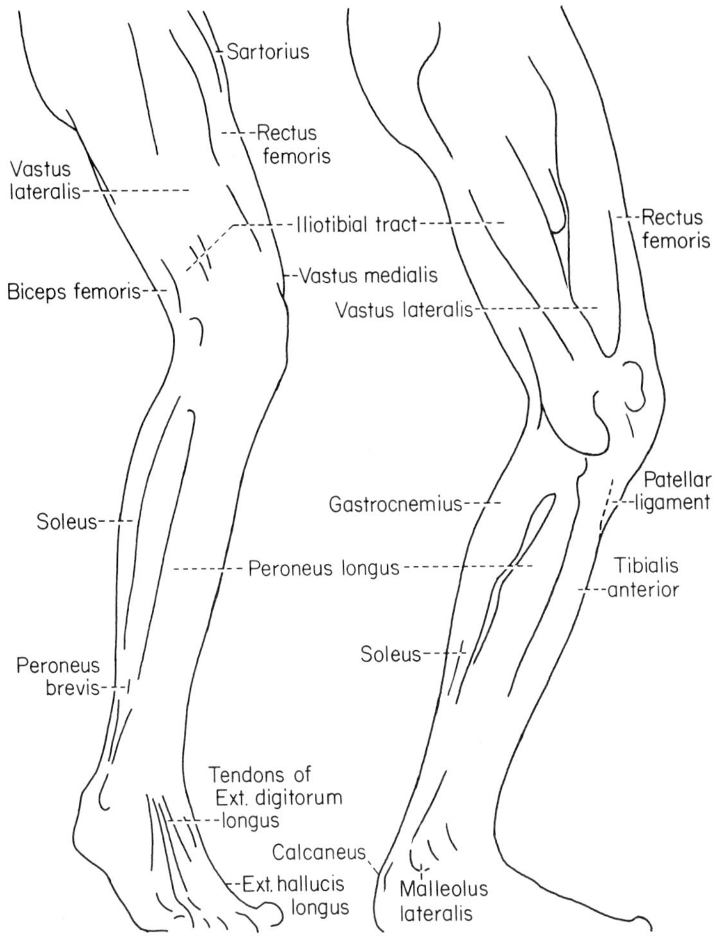

Figure 72 (♂ 40)

These right legs belong to a subject who, in his youth, participated in hiking, track and horseback riding. By standing with bent knees the knee extensors can be demonstrated. Additional raising on the toes brings the calf muscles into action. Attempted outward rotation in the hip (against friction of the floor) and eversion in the foot add even more muscle contours.

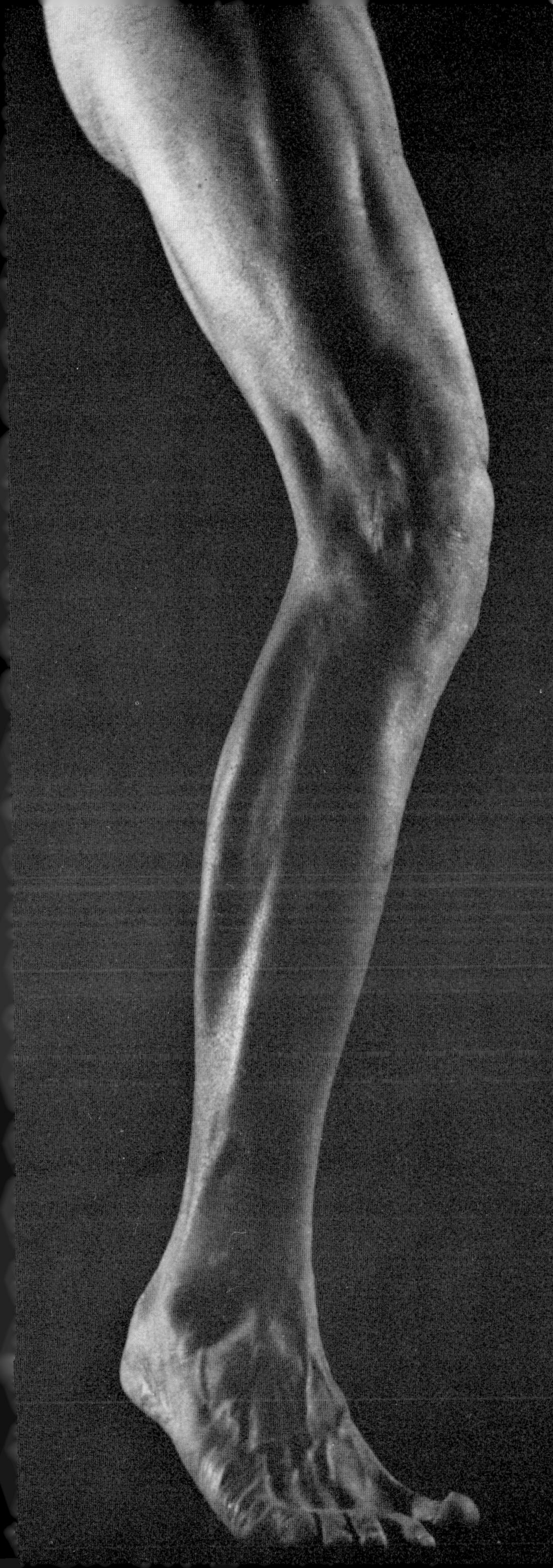
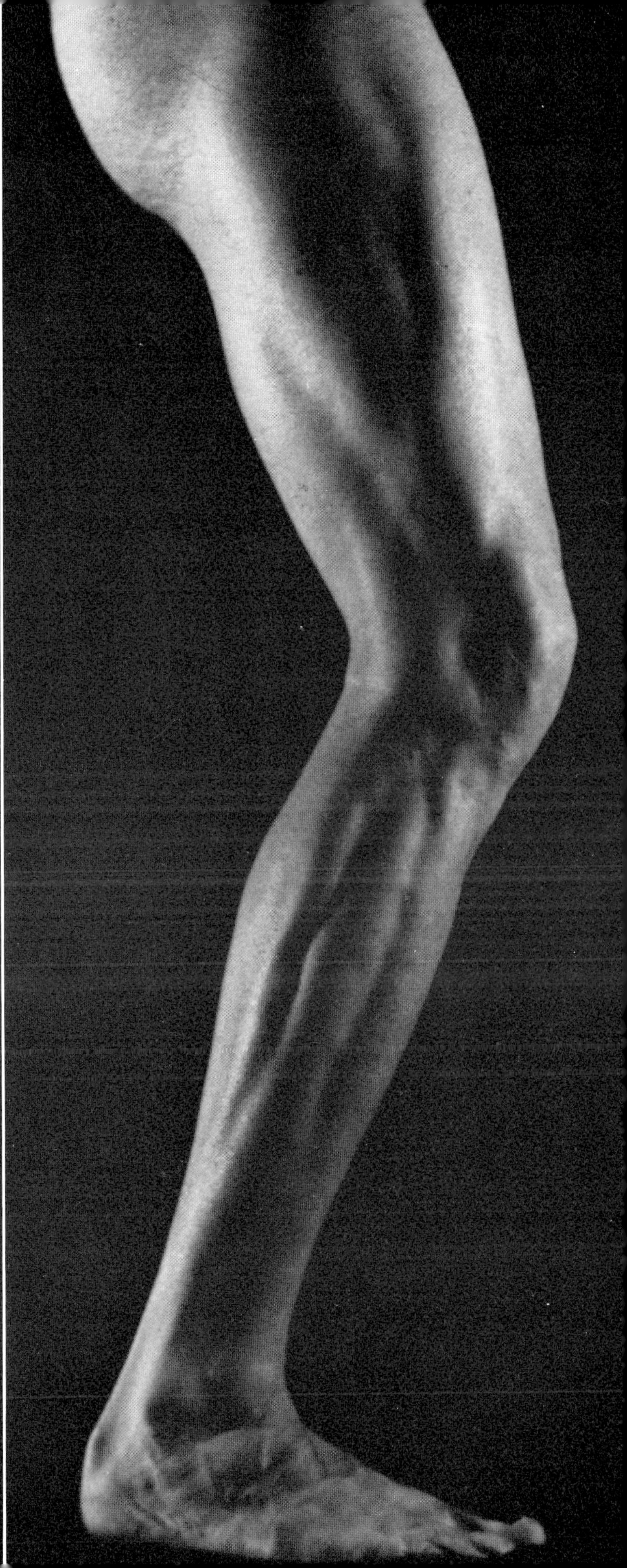

Figure 73 (♂ 28)

This subject shows smooth muscle contours. His muscles were well developed but he was unable to demonstrate the deep furrows, and separation of muscles as in Figure 72. His skin (adipose tissue included) was thicker and he has been less active than the previous subject.

In a flexed knee the tendons of the biceps femoris and semitendinosus-membranosis stand out. The space between the ends of these two tendons (called the popliteal space) is often filled with fatty tissue.

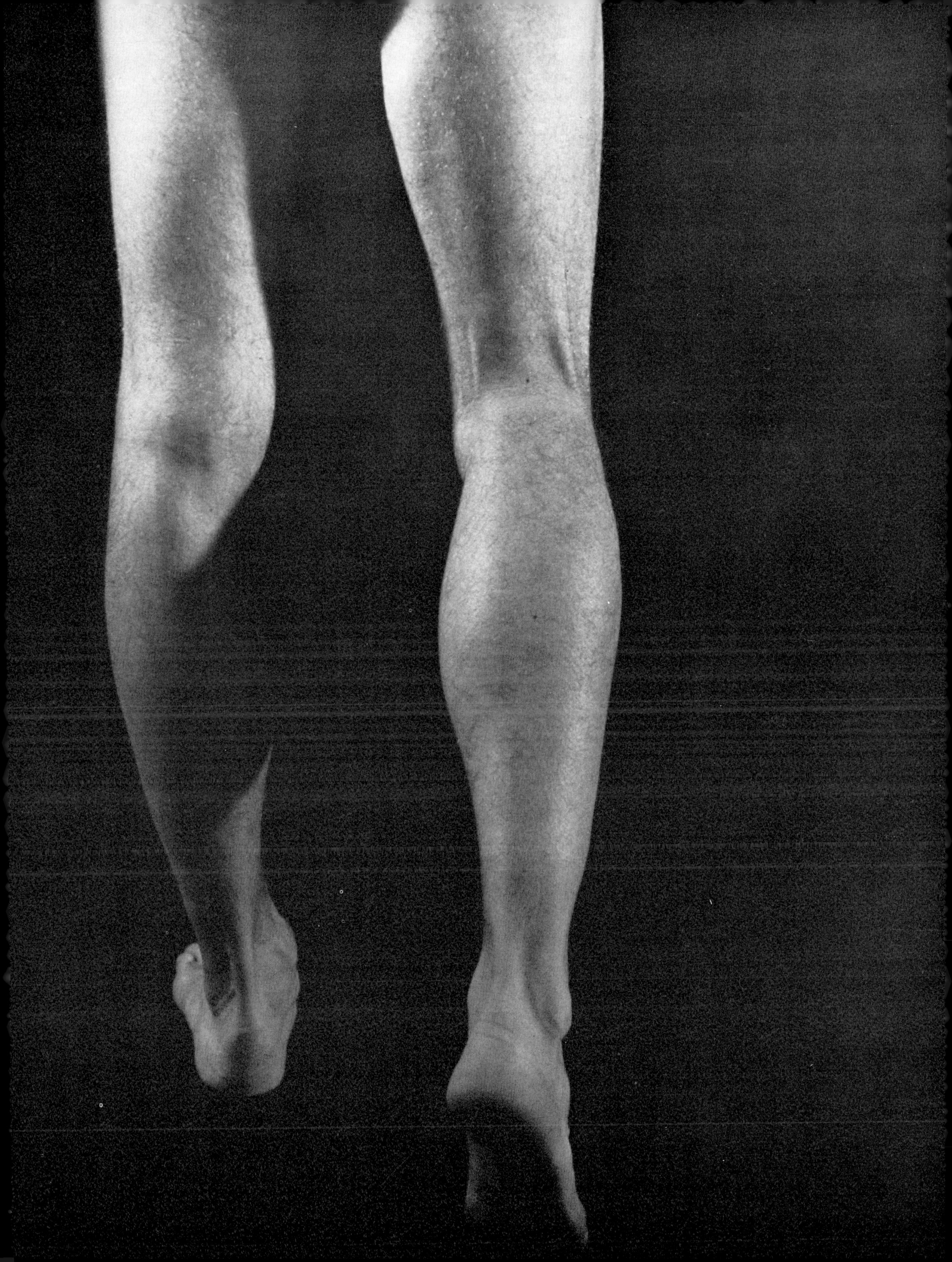

Figure 74 (♂ 8)

In this young subject the skin was thin (no fat). The muscles are smooth and not separated. Comparison of Figures 72, 73 and 74 shows the wide variety in muscle definition. This is not all due to activity or adiposity (see Fig. 75), but may have some relationship to age and maybe (natural) dehydration with age.

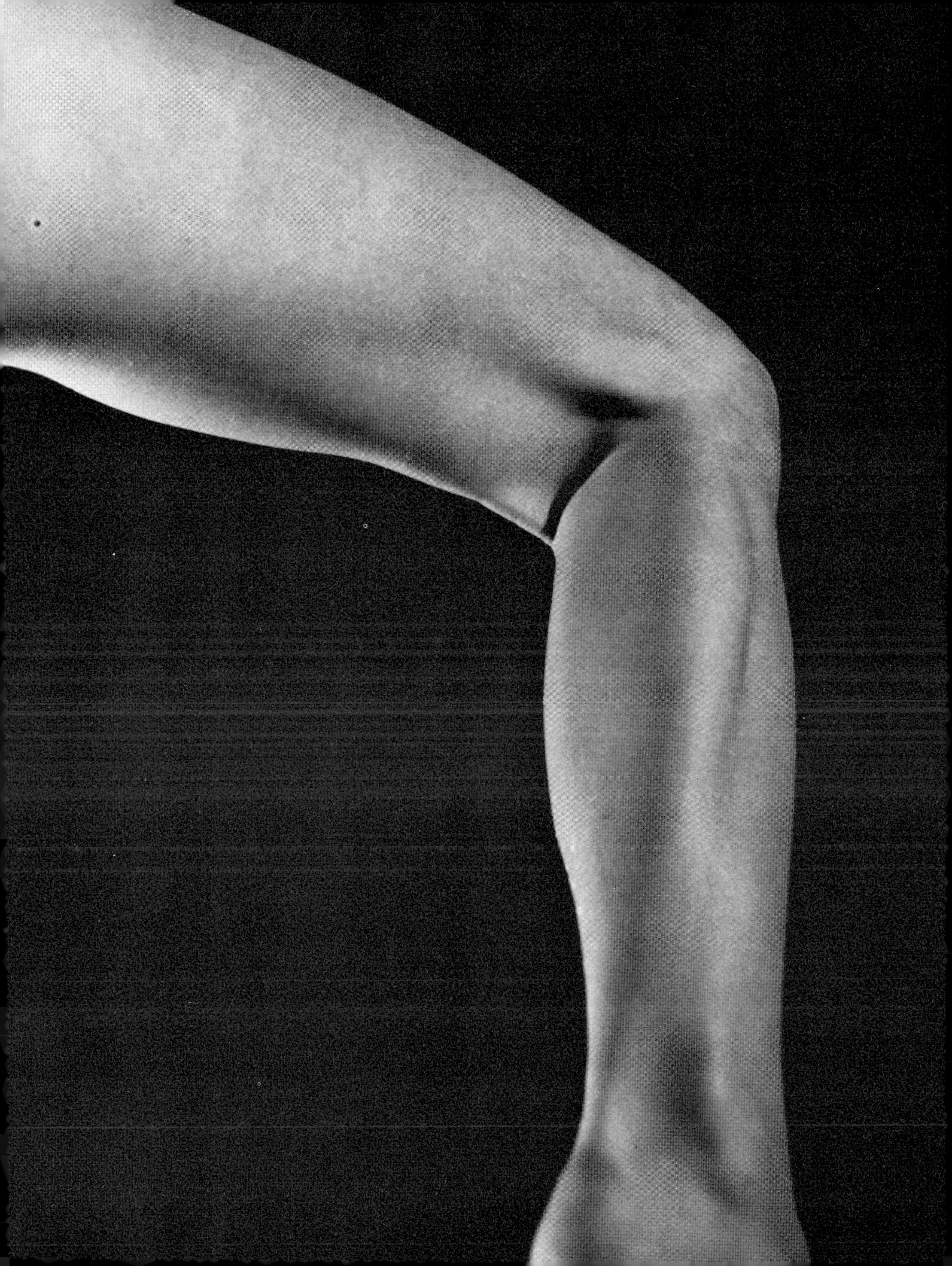

Figure 75 (♀ 50)

This subject shows little muscle definition because of overlying fatty tissue. However, daily, since she was fourteen years of age, she participated in calisthenics and rode a bicycle for nine miles (uphill). The fat was hard and firm.

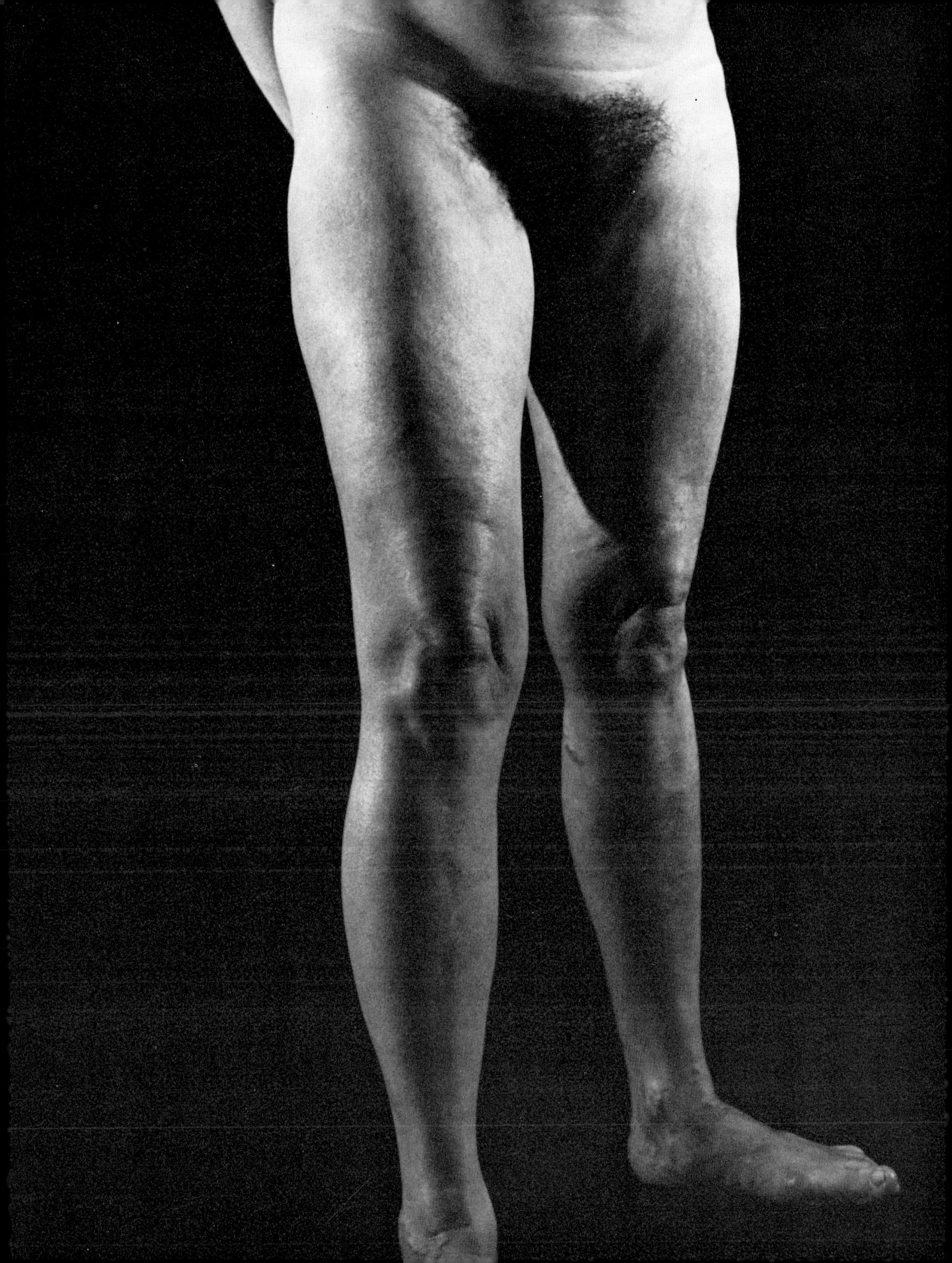

Figure 76 (♀ 12)

In a standing position the knee can be "locked" thus making the quadriceps unnecessary for maintaining the extended position. Tension on the iliotibial tract is maintained by the contraction (static) of the tensor fasciae latae. With a straight knee this tract acts as an extensor on the knee. The subject was asked to alternately drop (left) and raise (right) the patella. Aside from the shift (upwards) of the patella, very little difference in contours of the leg resulted in this thin, young subject.

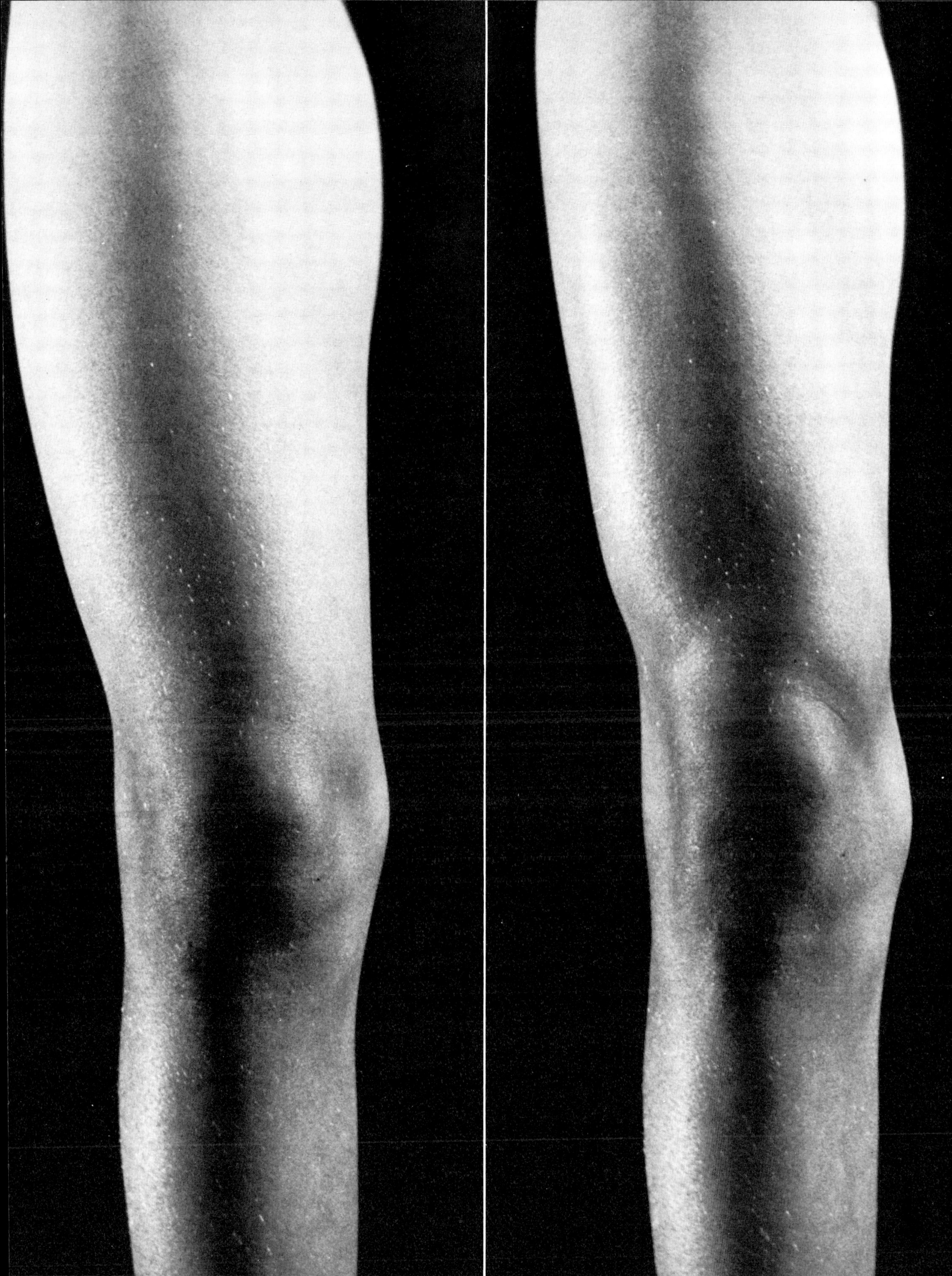

Figure 77 (♀ **11**)

Extreme passive flexion in the knee is generally limited by the bulk of the hamstrings. Active flexion is limited by the extent to which the hamstrings can shorten against the stretching of the antagonistic quadriceps.

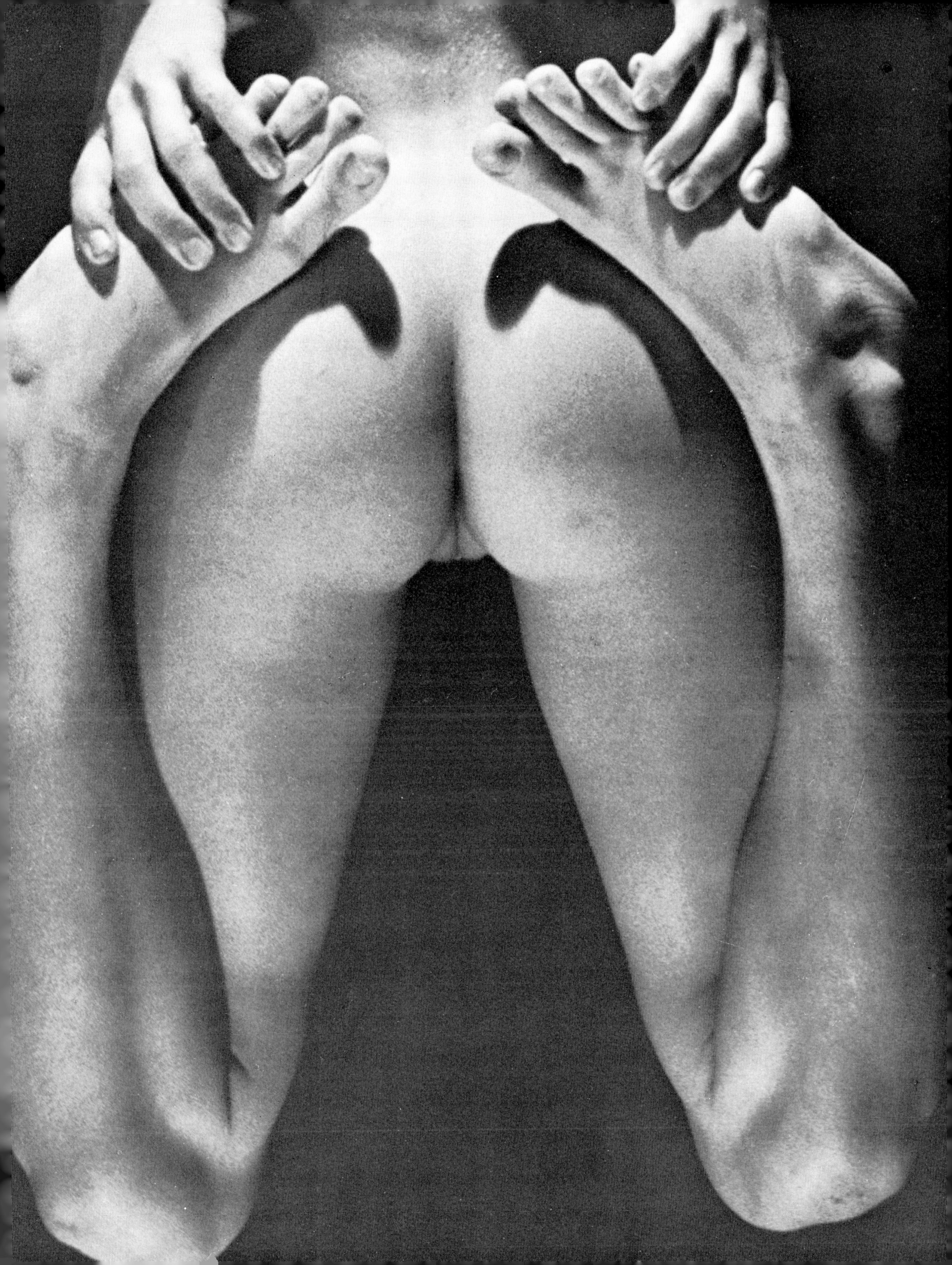

Figure 78 (♀ 6)

This little girl stands with the knees hyperextended (genu recurvatum, left). In addition, the abdominal wall is sagging, the lumbar part of the spine is hollow (lordotic) and the upper part of the spine is convex (kyphotic) in compensation. The only instruction given was, "Pull your belly in," which resulted in (middle) a backward tilting of the pelvis, and a straightening out of the knees and the lower and upper back. The front view of the first (uncorrected) pose shows this child to be slightly bow-legged (genu varum).

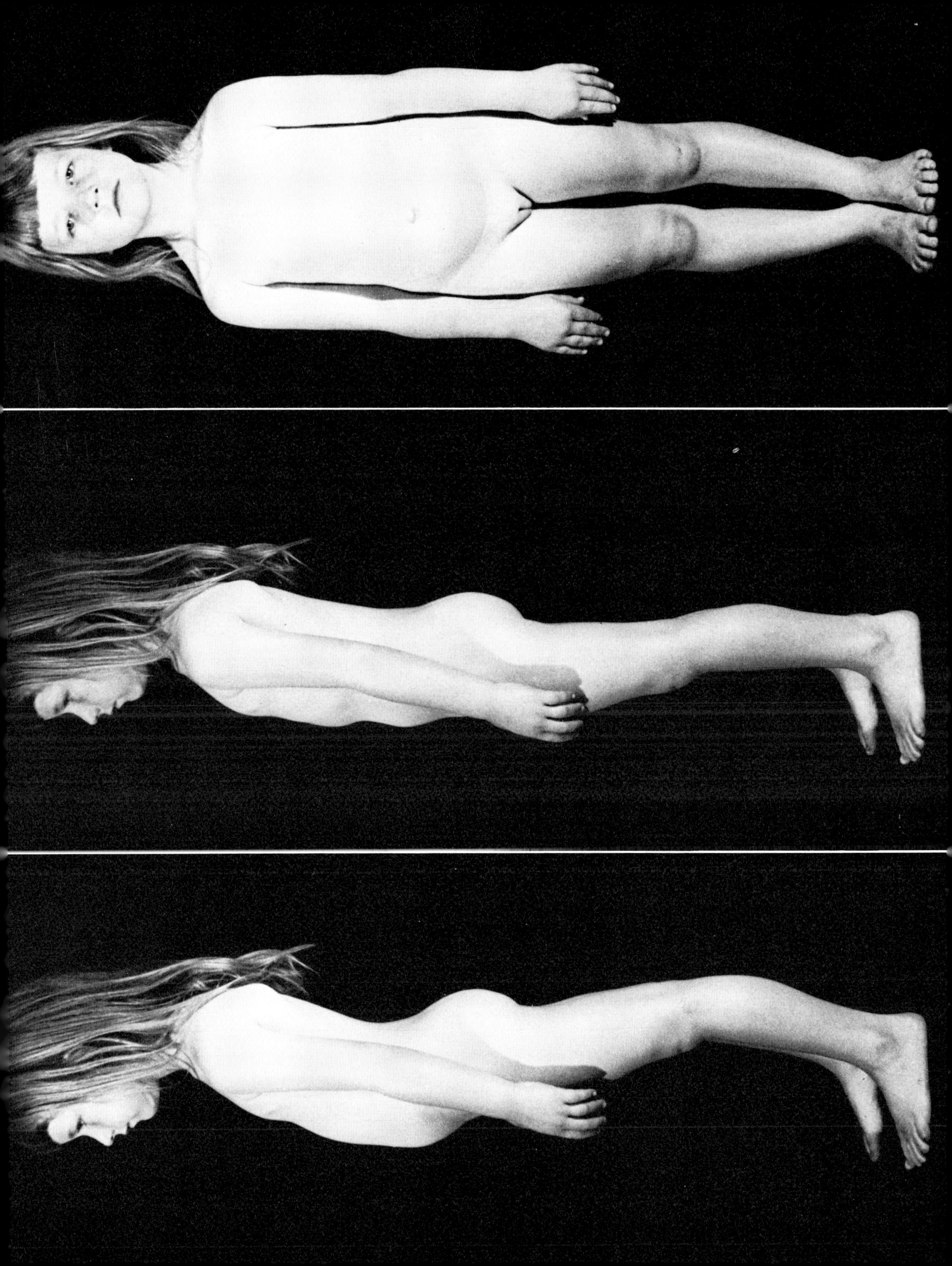

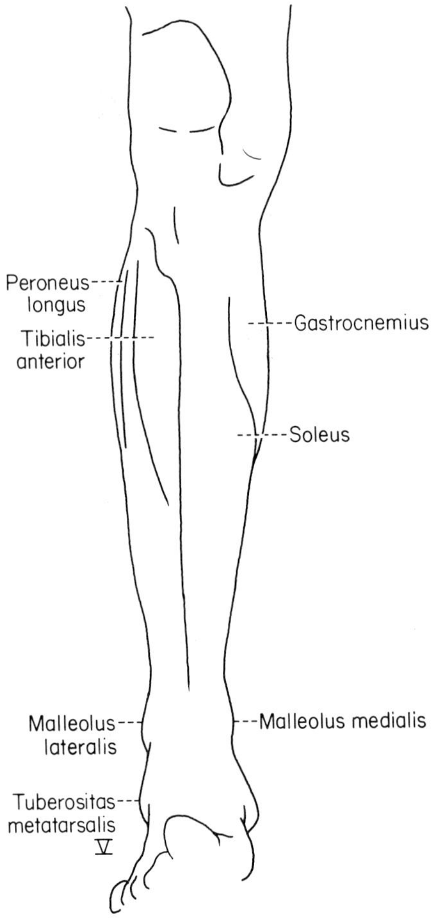

Figure 79 (♂ 40)

The same subject as shown in Figure 72. While bearing the weight on the heel of the foot (with bent knee) an attempt is made to lift the front of the foot and the toes off the floor (dorsal flexion in ankle and foot) to bring out the contours of the tibialis anterior just laterally from the sharp edge of the tibia.

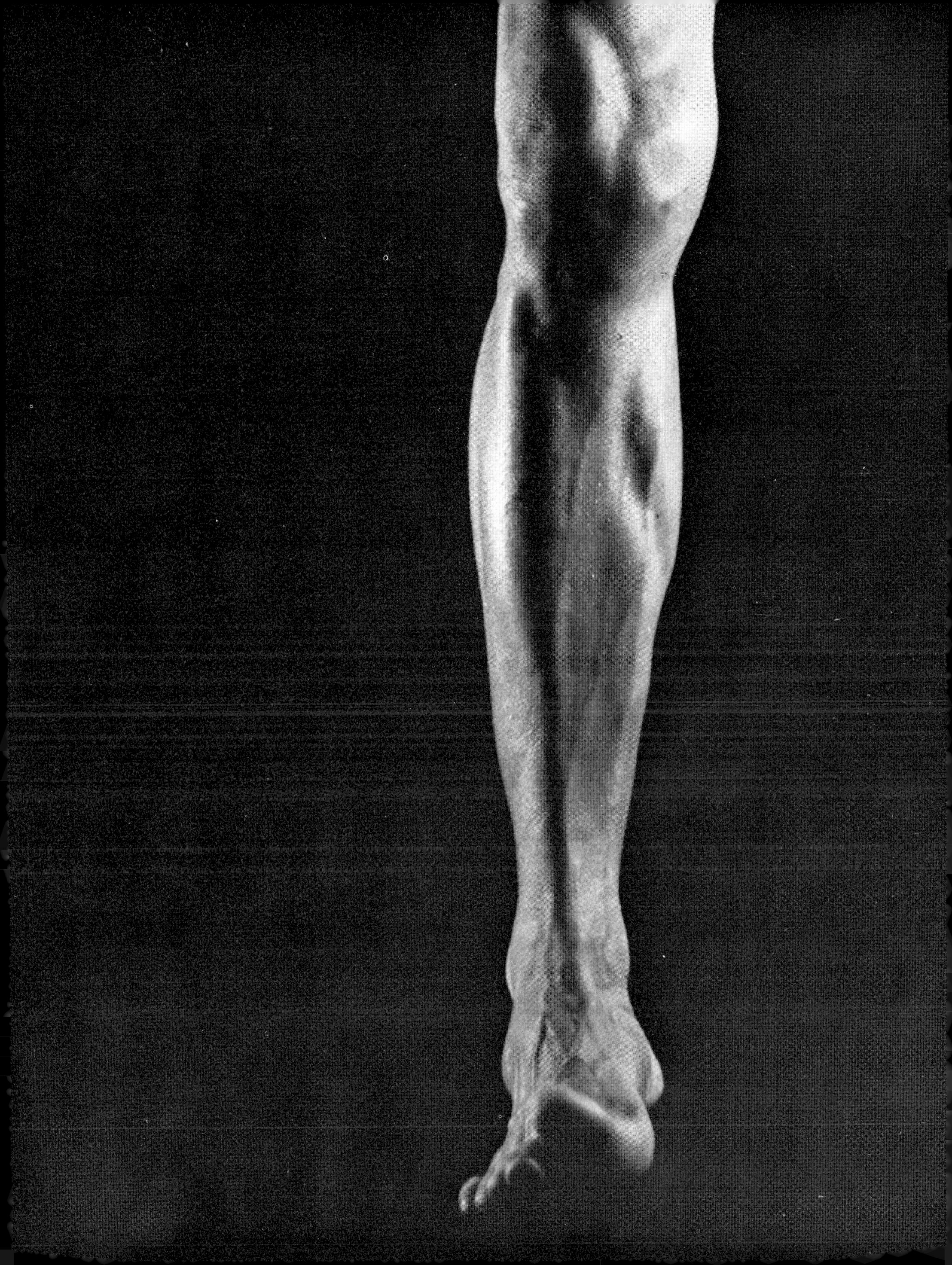

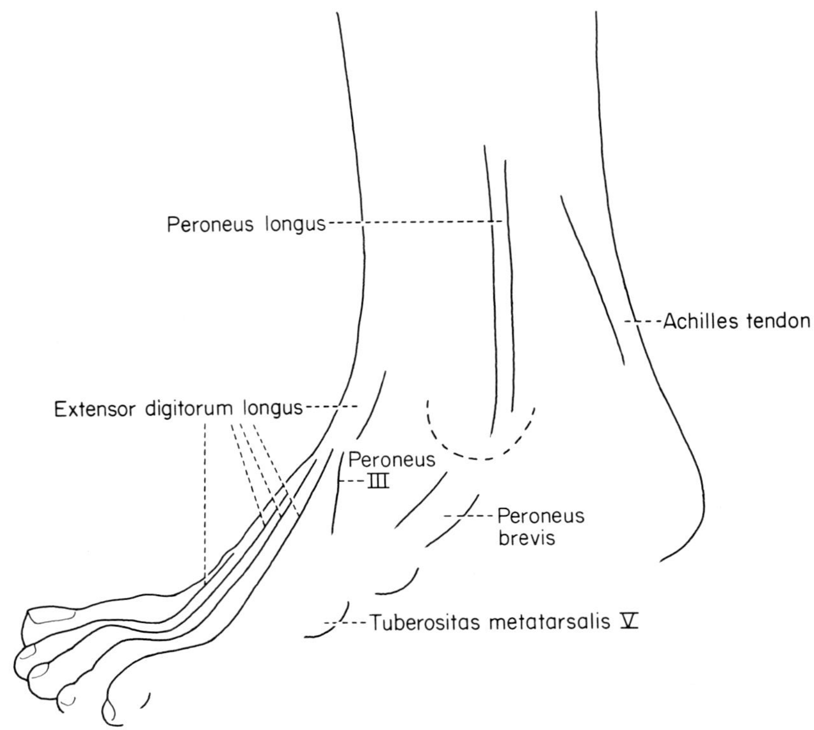

Figure 80 (♀ 10)

In a sitting position and with a slight resistance of the lateral edge of the foot against the floor, the heel was raised. This action brought the tendons of the extensor digitorum longus clearly into view. The peroneus longus is visible in its course behind the lateral malleolus.

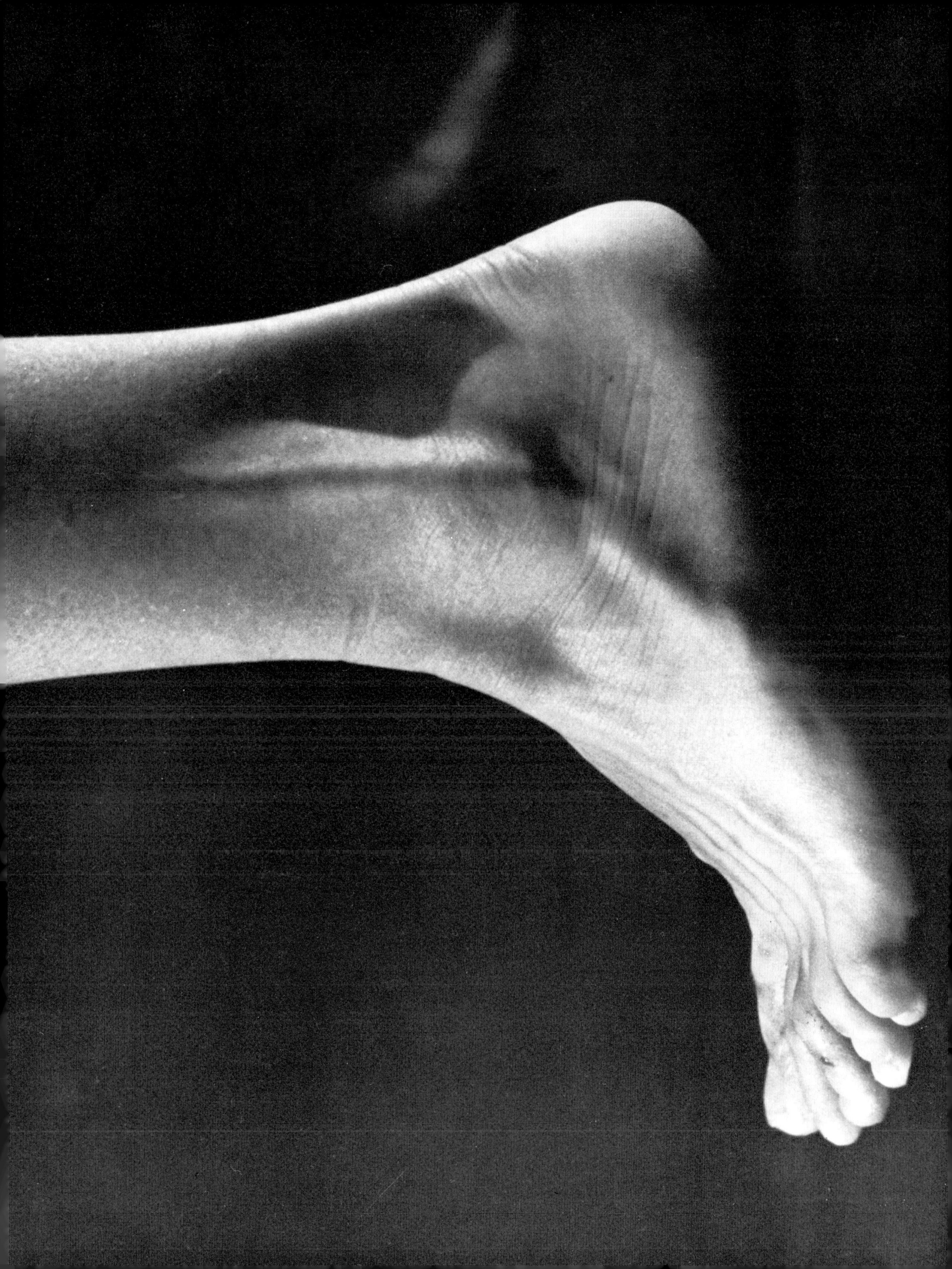

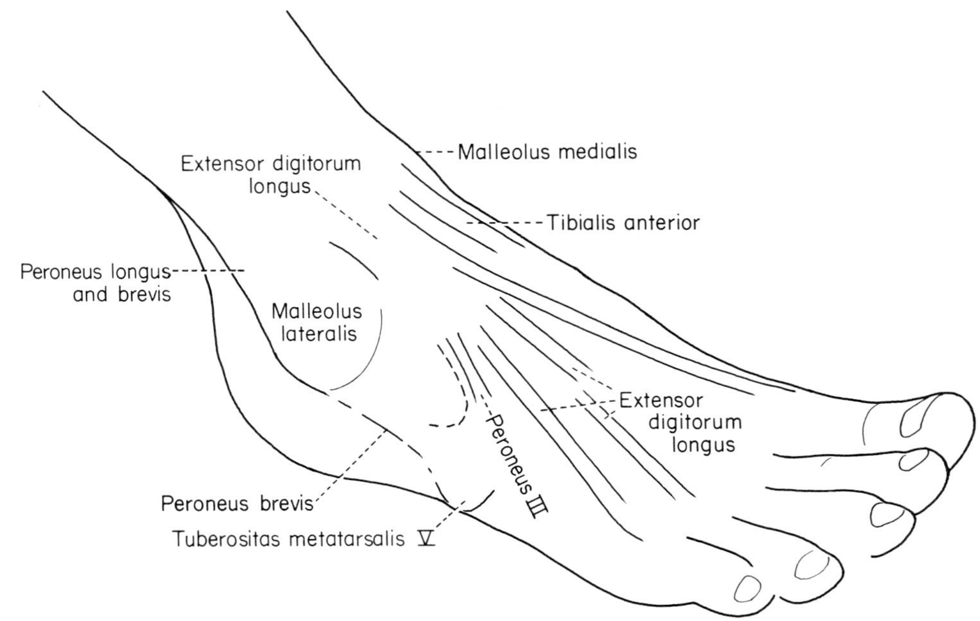

Figure 81 (♀ 11)

With little resistance from the floor against the heel the front of the foot is raised and the toes extended to make the tendons of the extensor digitorum longus visible. The peroneus longus, brevis and tertius are all visible.

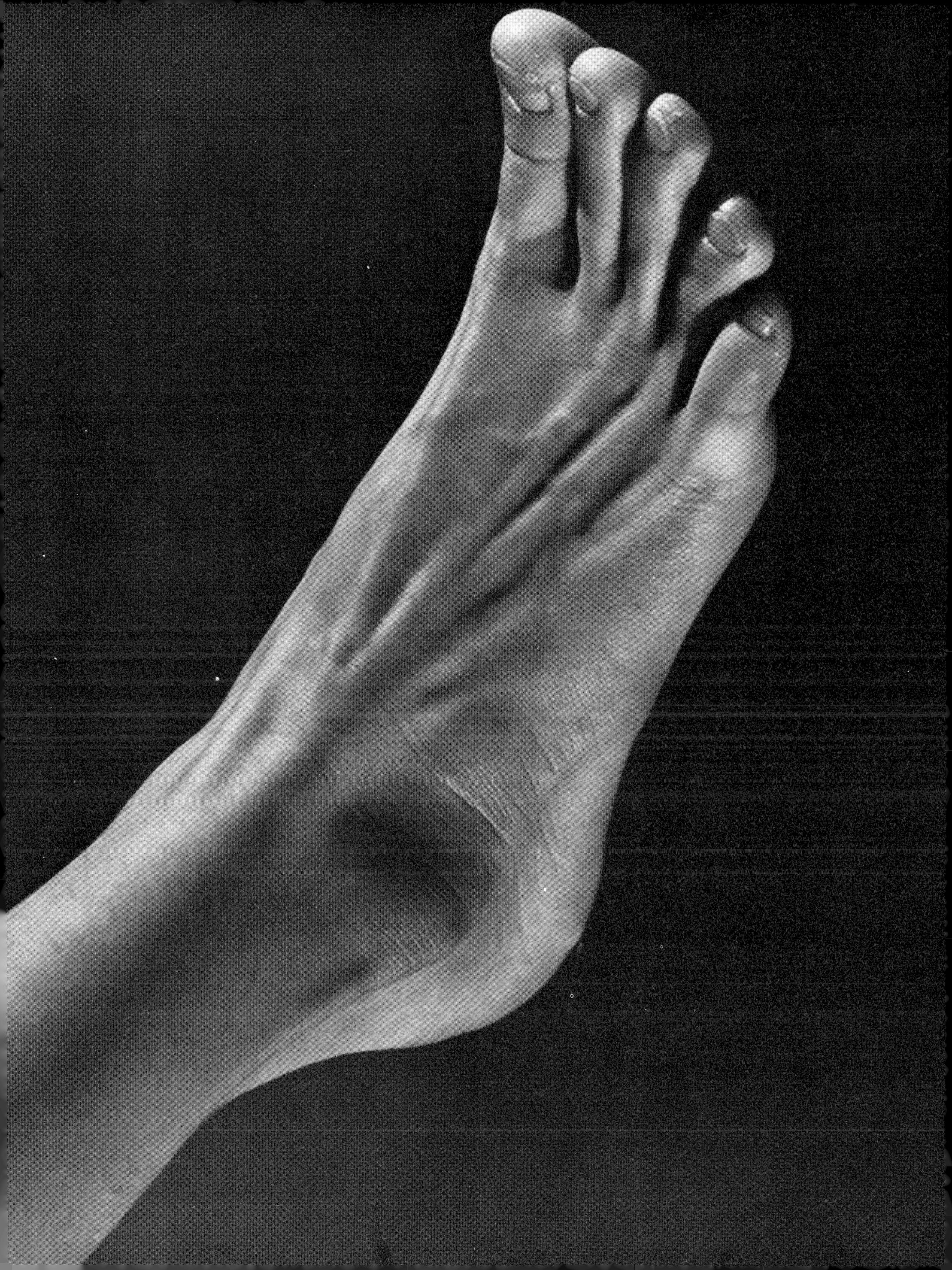

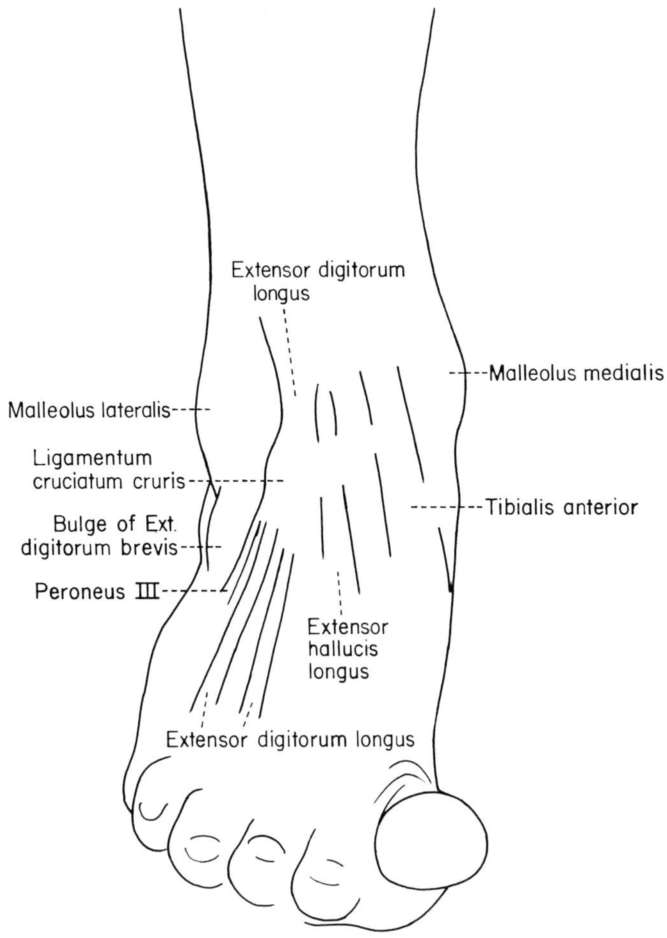

Figure 82 (♀ 8)

The place where the extensor digitorum longus crosses underneath the ligamentum cruciatum cruris can be seen.

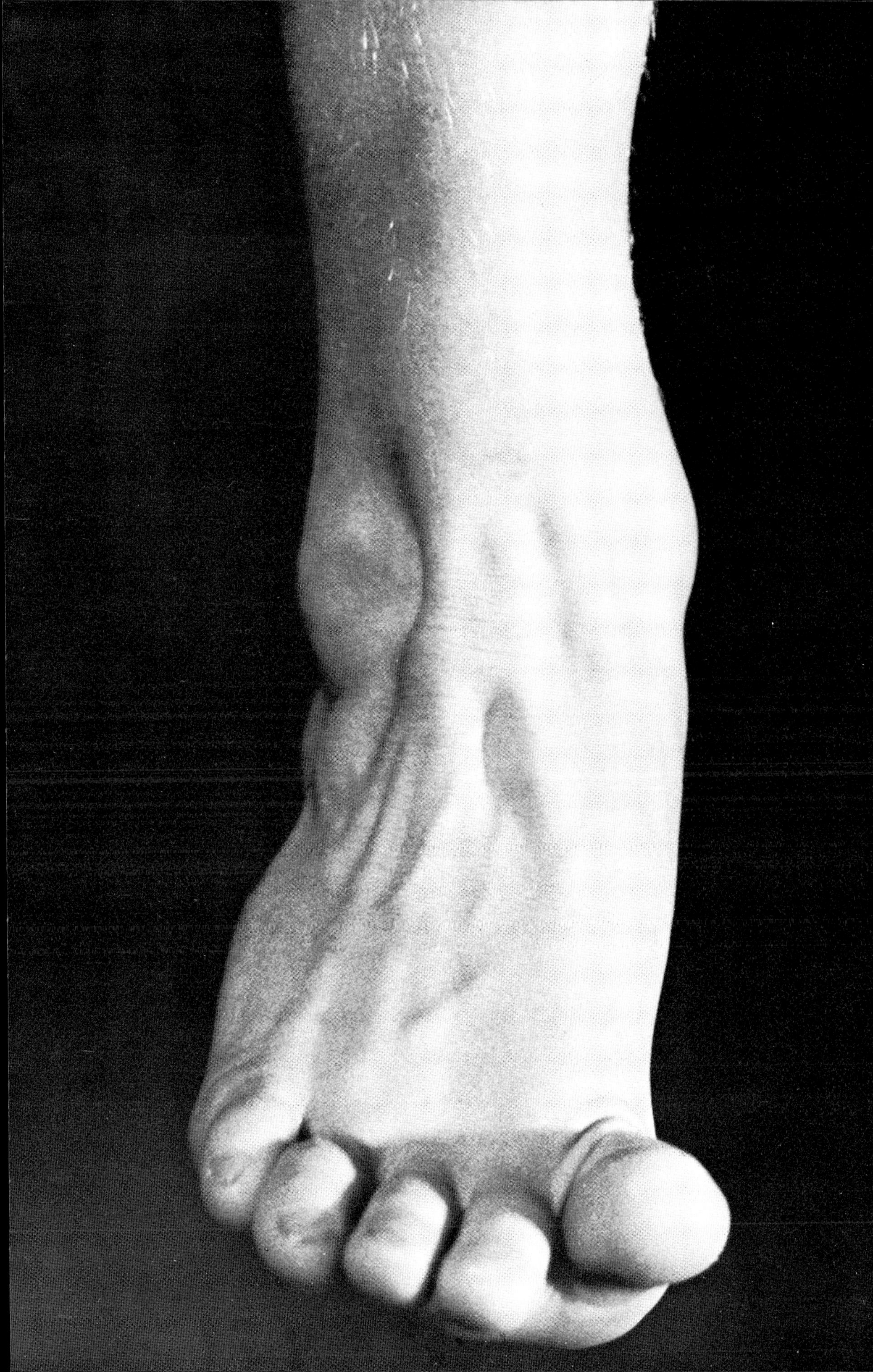

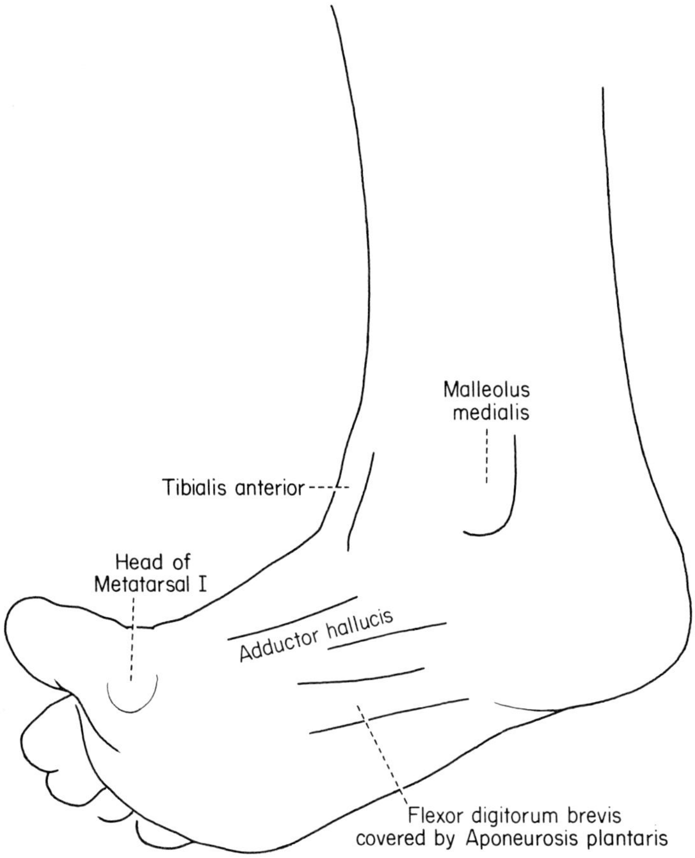

Figure 83 (♂ 6)

While the lower leg hangs freely over the edge of a table the front of the foot is supinated (the foot sole is twisted medial-ward) and lifted. The contours of the foot-sole muscles and ligaments are generally obscured by the aponeurosis plantaris, the fat padding and the thickness of the skin.

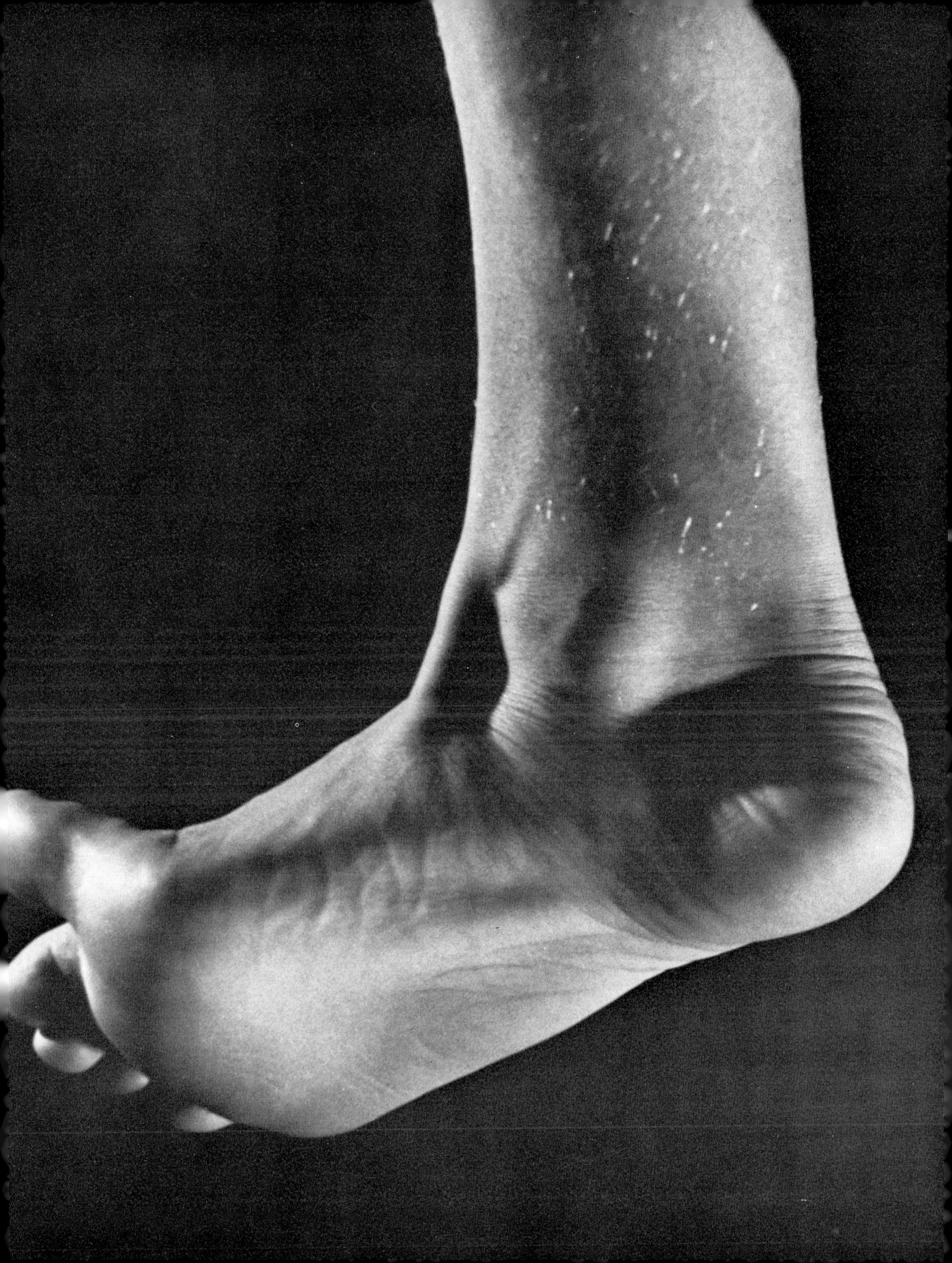

Figure 84 (♂ 2 months)

The subcutaneous fat layer obscures the muscles, tendons and bony landmarks in most infants (compare with Figs. 80 and 81).

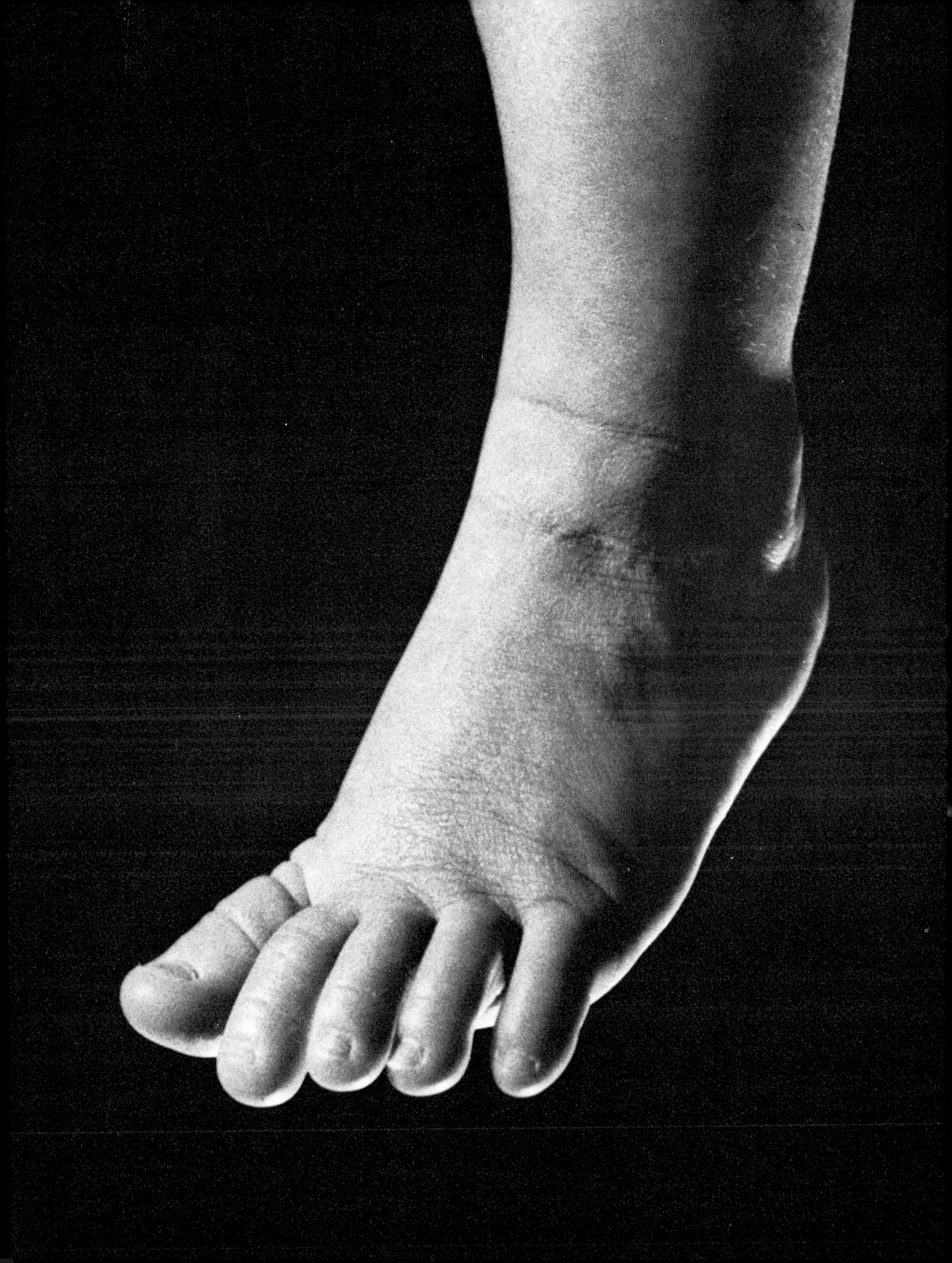

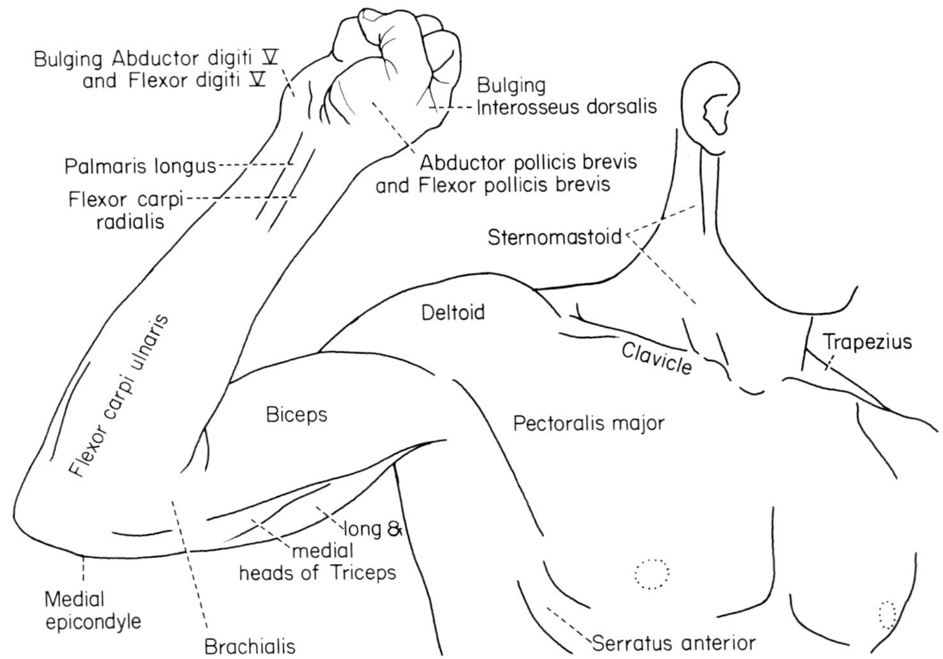

Figure 85 (♂ 26)

In this subject the relationship between neck, shoulder, chest and arm muscles is clearly visible. Note the lateral end of the pectoralis major which runs over the biceps and disappears underneath the anterior deltoid. The biceps muscle is in an elongated position since the insertion on the radius is moved away from the origin due to the pronation of the forearm.

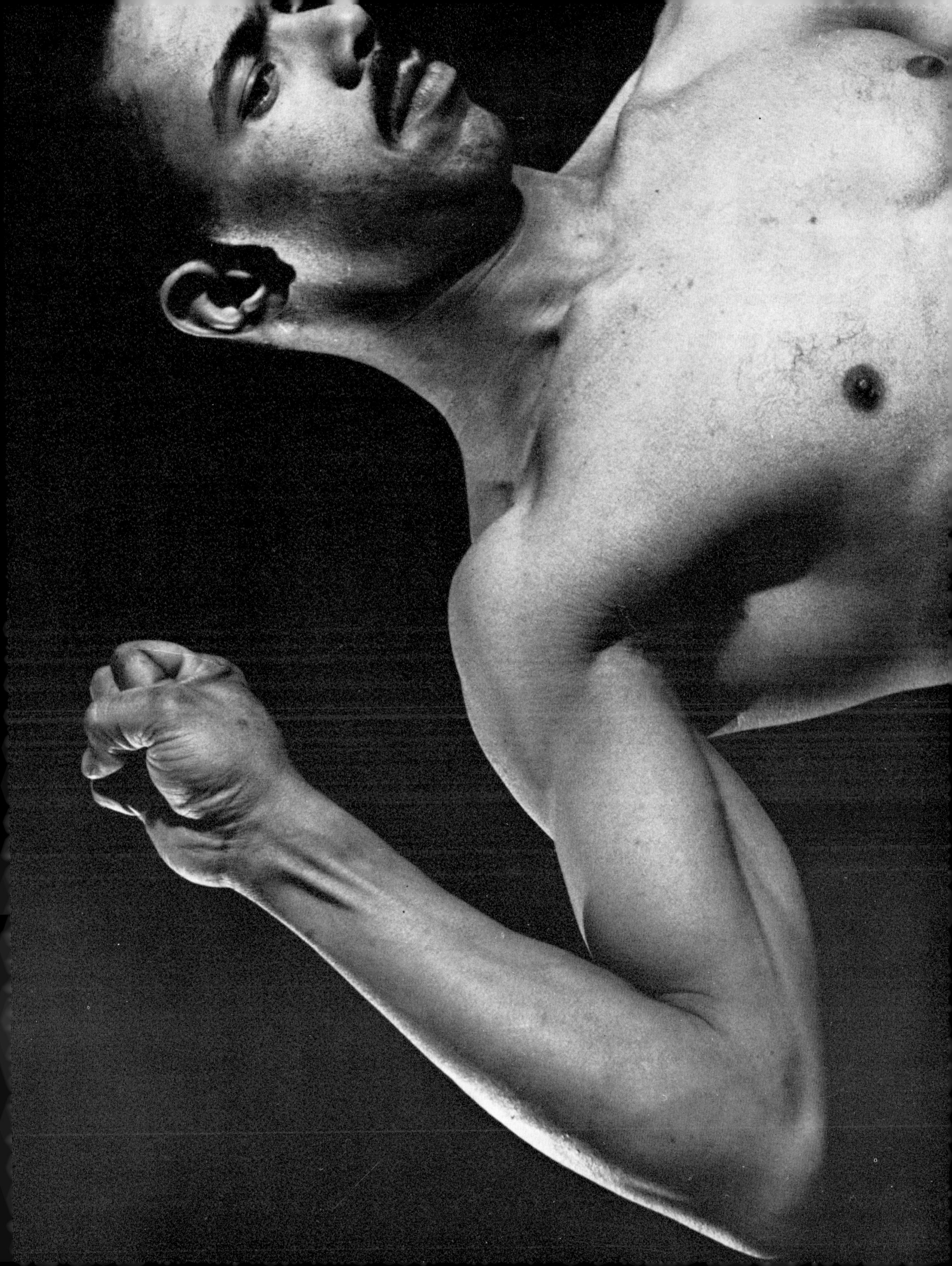

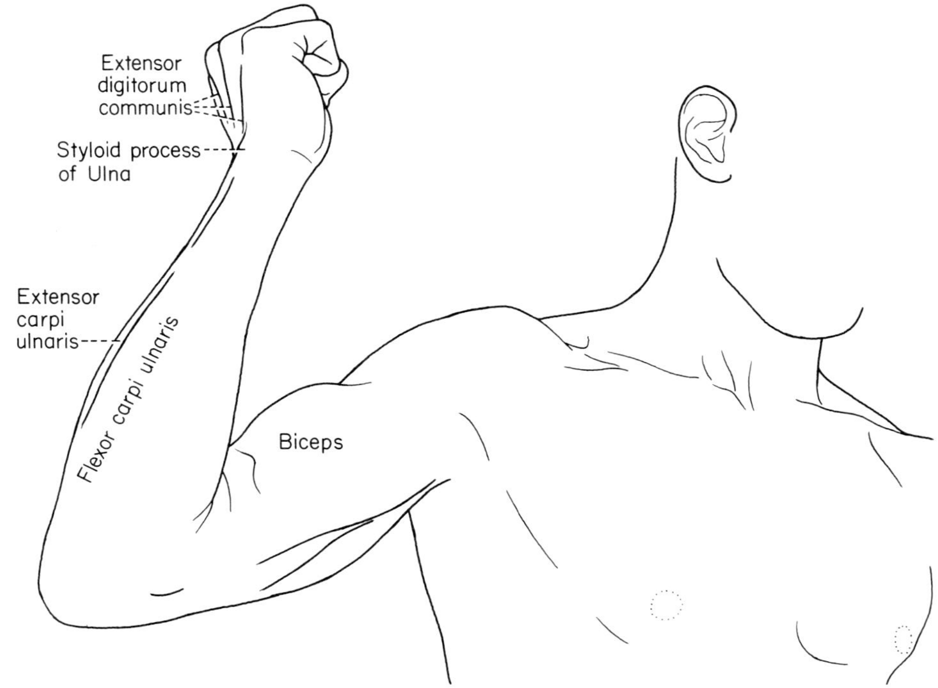

Figure 86 (♂ 26)

When the forearm is returned from the pronated position of Figure 85 and further supinated, the biceps shorten to form a more rounded bulge. In strong subjects, flexion in the elbow is often limited by the bulk of the biceps (compare with Fig. 87).

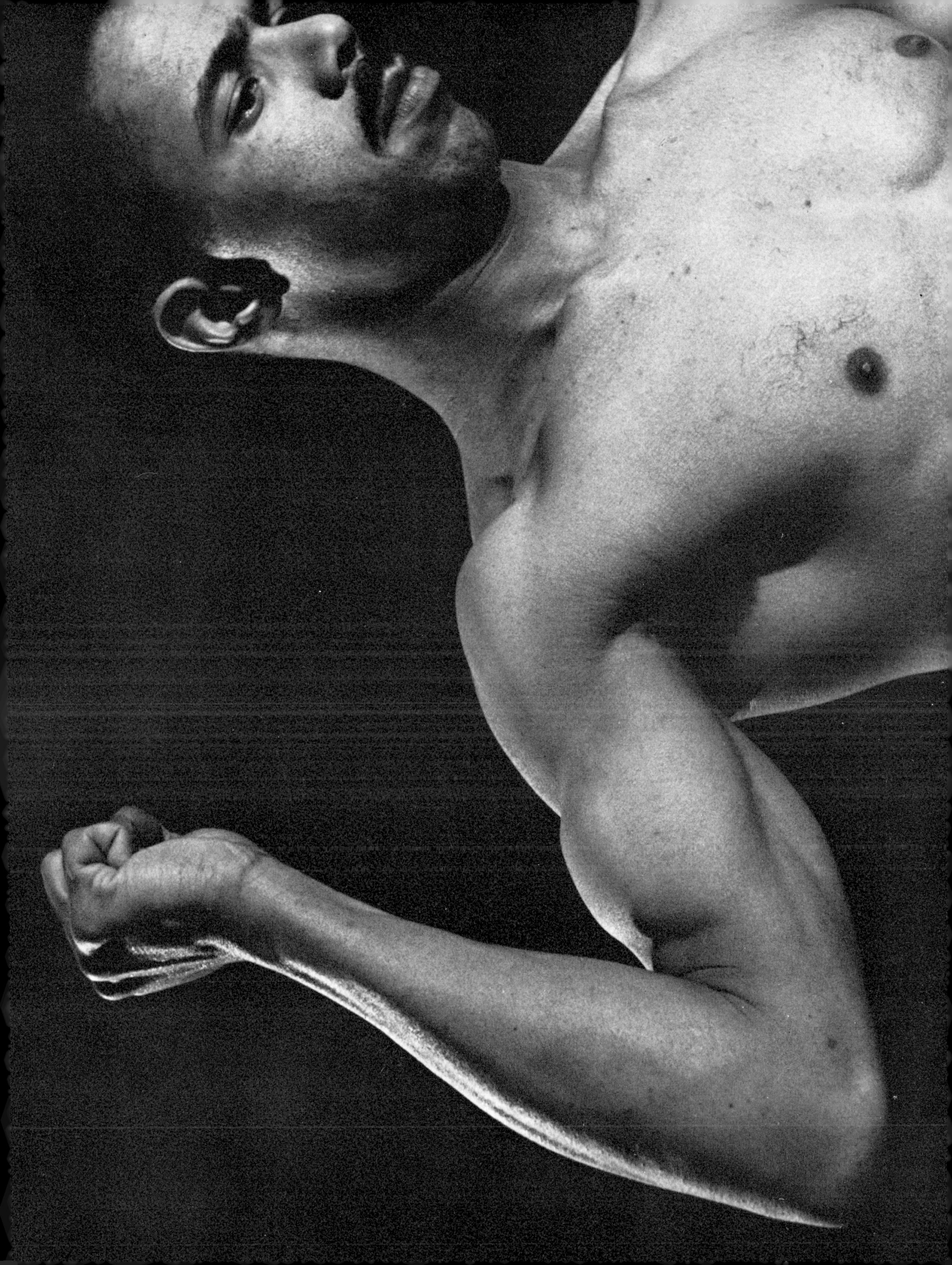

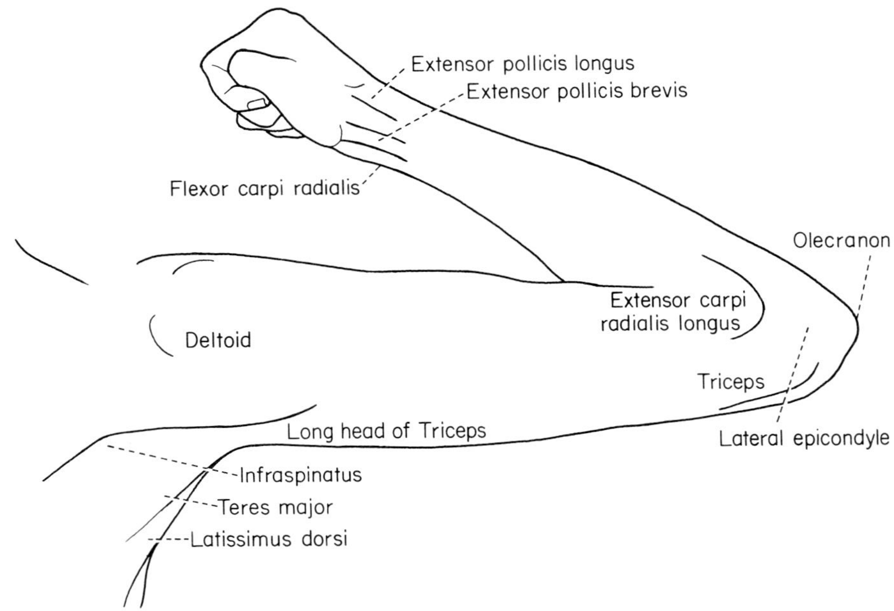

Figure 87 (♀ 24)

In contrast to Figures 85 and 86, this subject shows very little muscle definition. Subcutaneous fat was absent and strength was apparently normal. Strength may have little relation to muscle bulk or muscle definition.

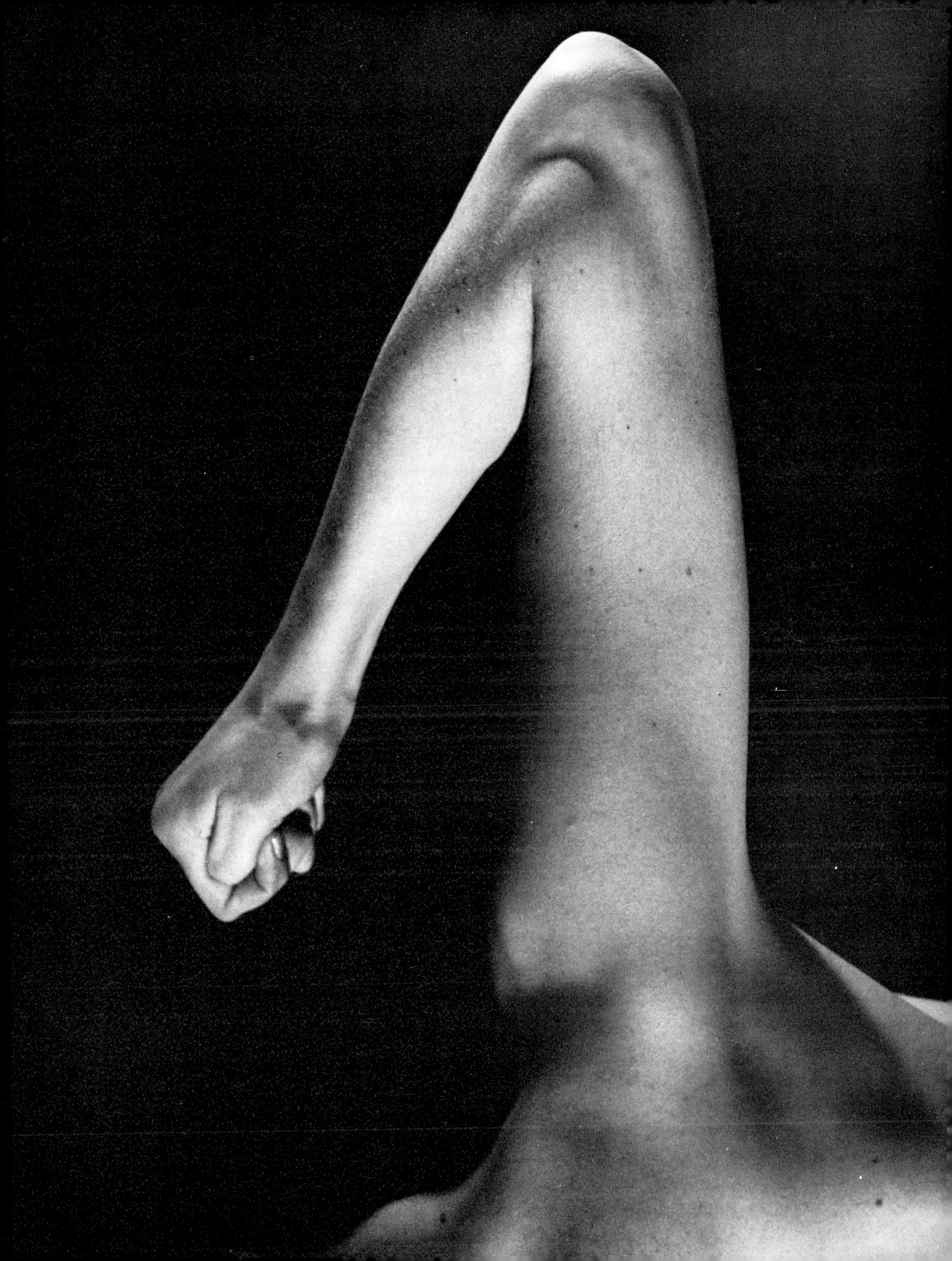

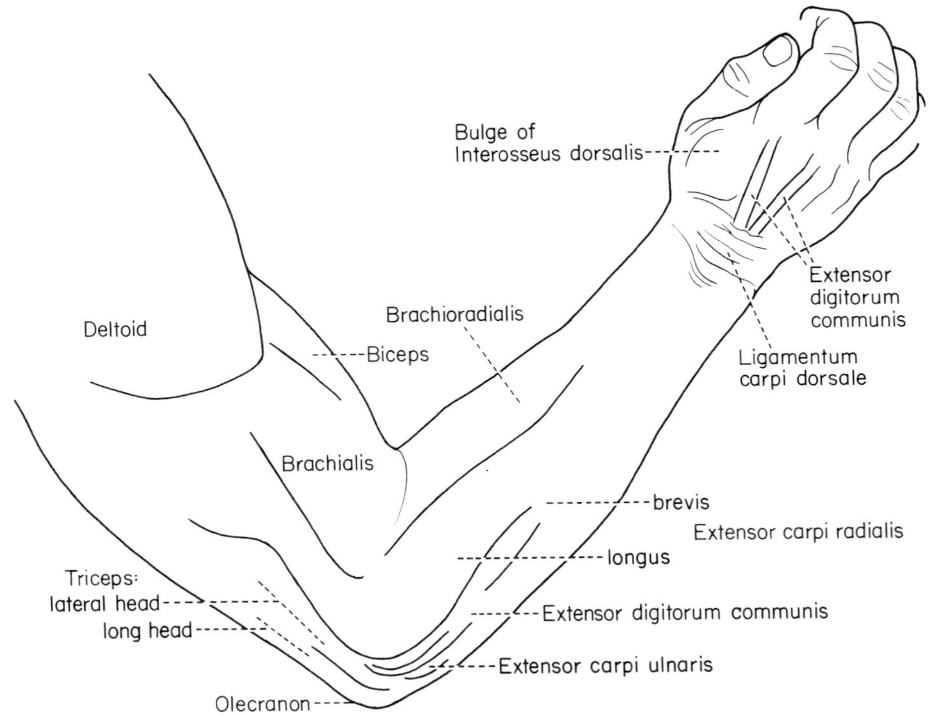

Figure 88 (♂ 38)

Static contraction of the antagonists kept the arm immobile while the muscle contours became visible. The subject has been active in lifting and carrying heavy weights most of his life. The skin is losing some of its resiliency.

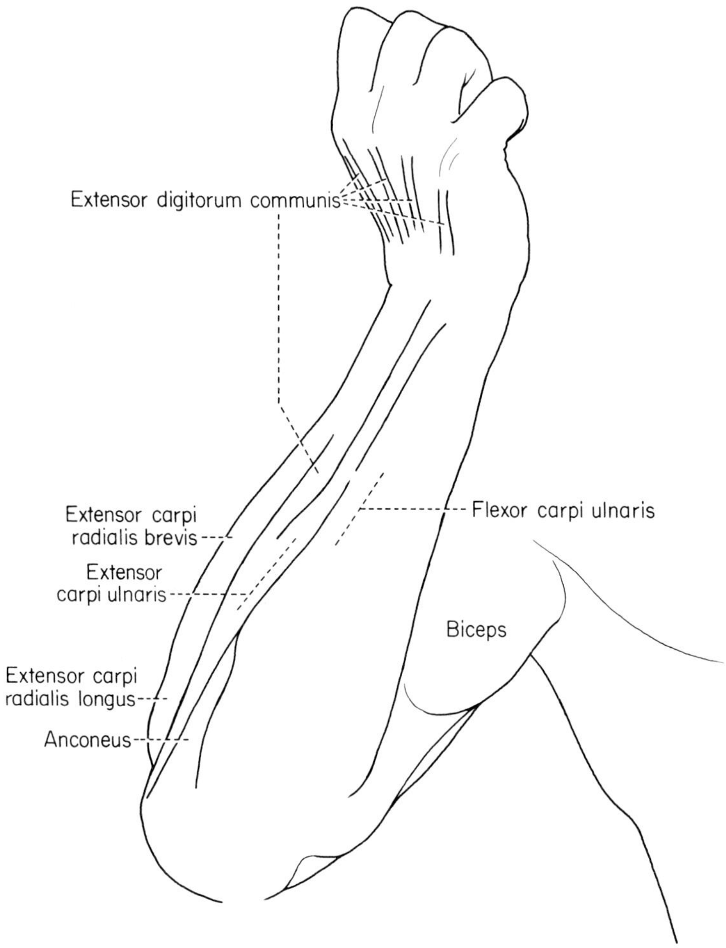

Figure 89 (♂ 38)

The same subject as shown in Figure 88 in a different static position.

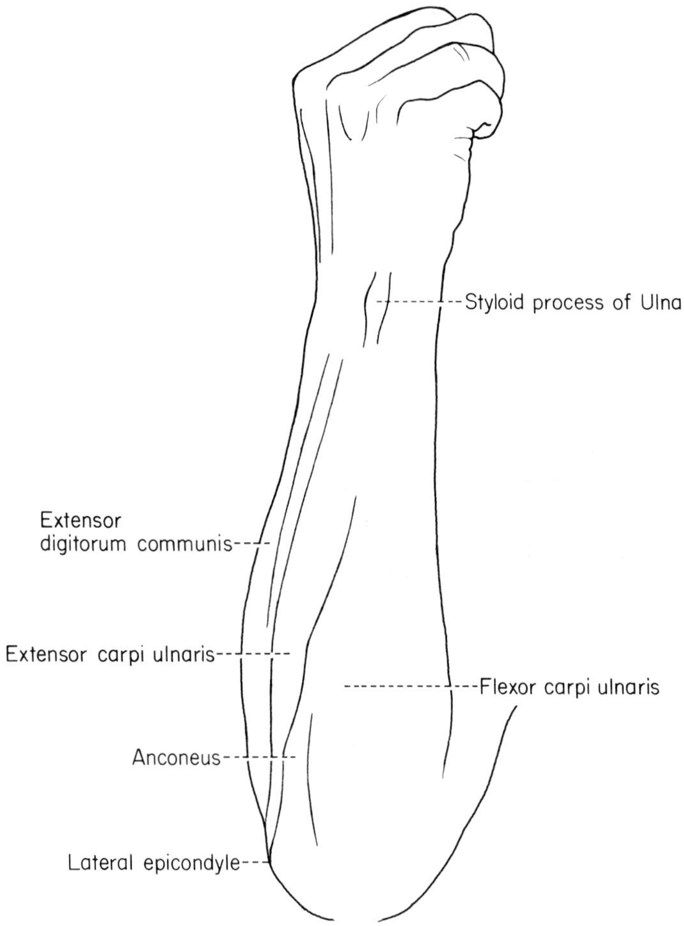

Figure 90 (♂ **38)**

The same subject as shown in Figure 88. Pinching the fingertips together in this tense, static position brought out the deep furrows between the forearm muscles.

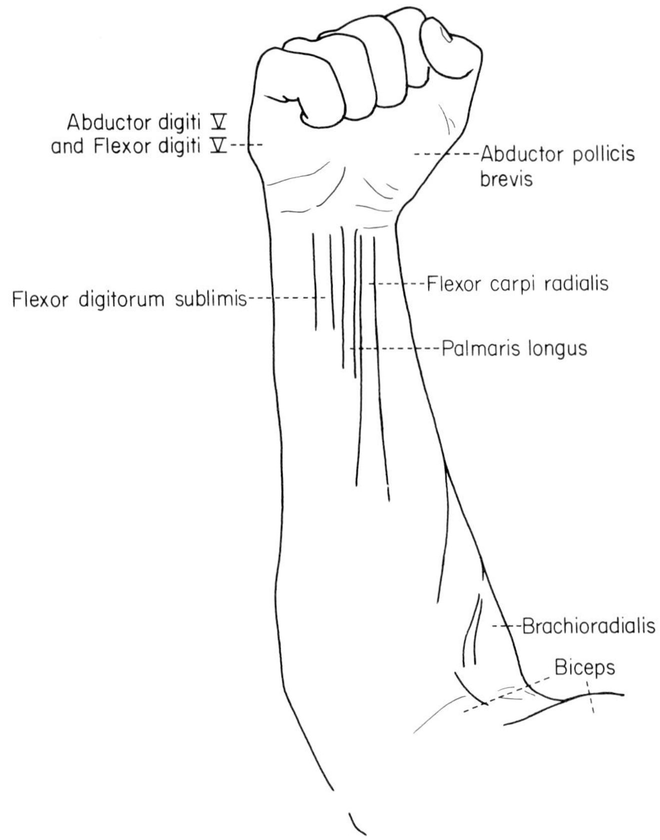

Figure 91 (♂ 36)

After squeezing a tennis ball with maximal force at a rate of thirty squeezes per minute for five minutes, the muscles became hard and large. The one-way valves in the vena basilica (basilic vein) and its branches are clearly seen.

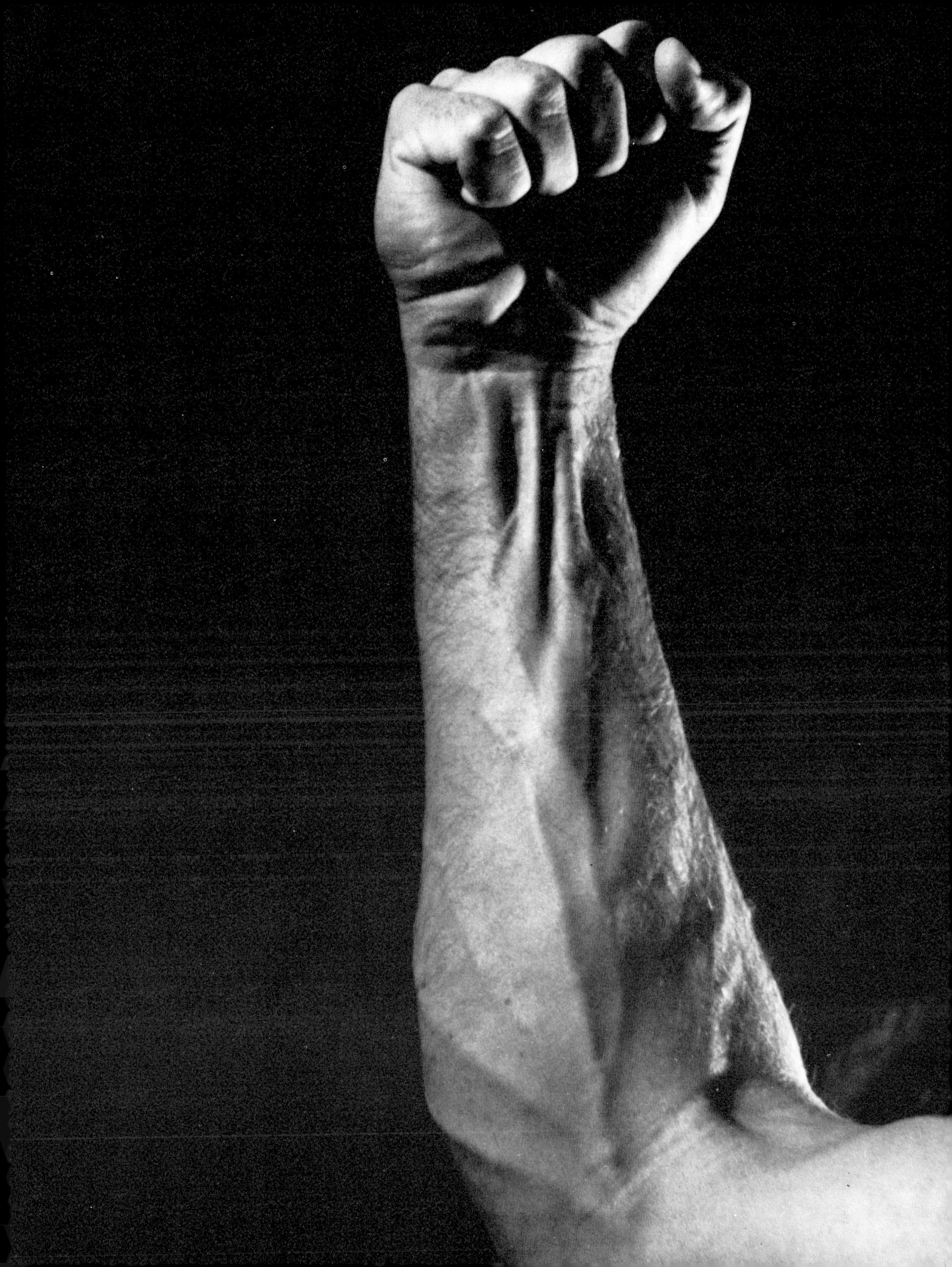

Figure 92 (♂ 72)

The same subject as shown in Figures 51 and 70. Although the overlying loose skin obscures most of the muscle contours (a twenty-pound weight is held in the hand) of the elbow flexors, it was possible to detect good development by palpation.

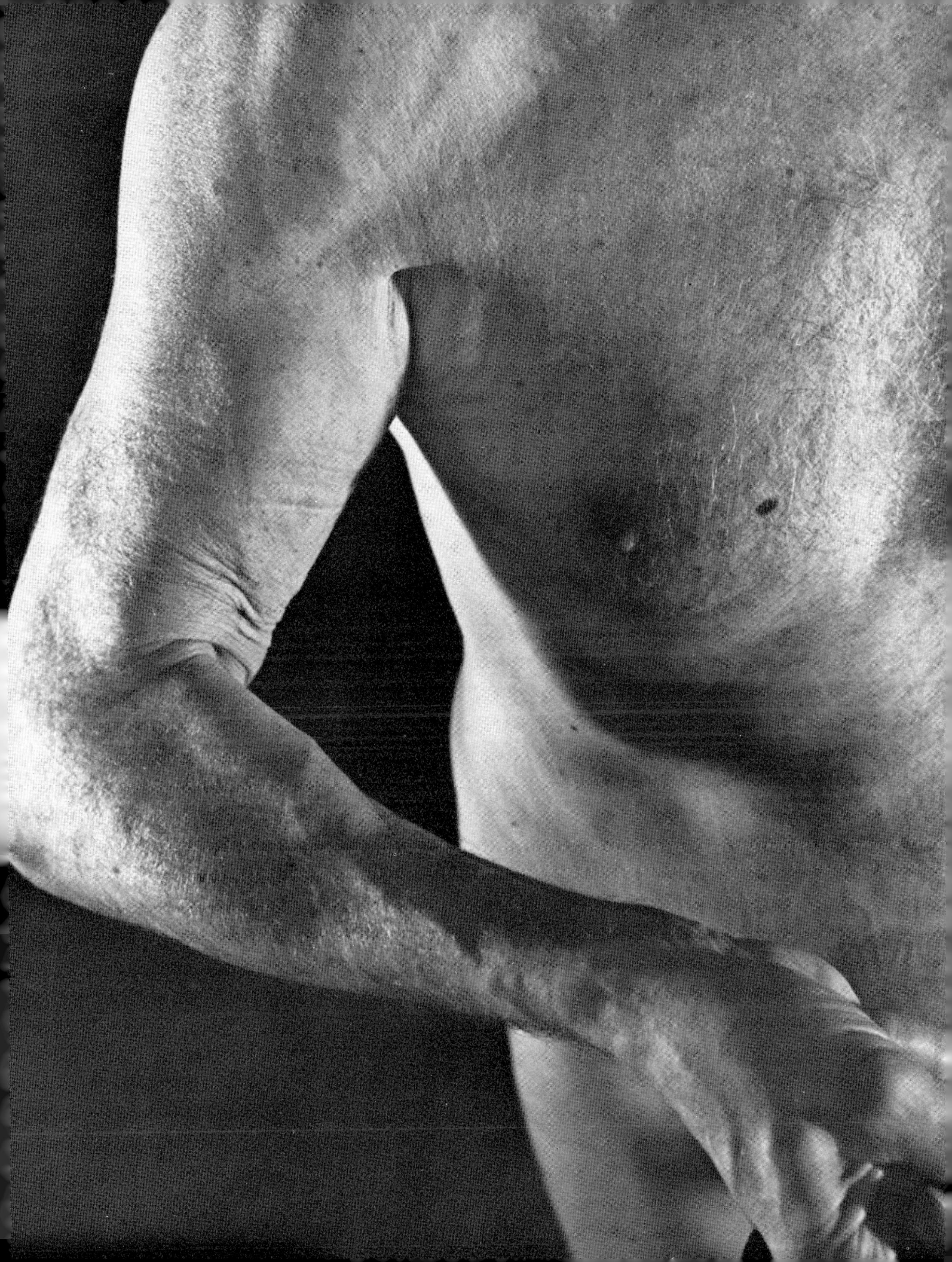

Figure 93 (♂ 18)

In contrast to the man shown in Figure 92 the arm of this young adult shows a smooth skin. The vena cephalica is visible on the back of the wrist and hand. The styloid process of the ulna, and the olecranon are very marked in this tall, thin subject. Fingers show some hyperextension in the joints.

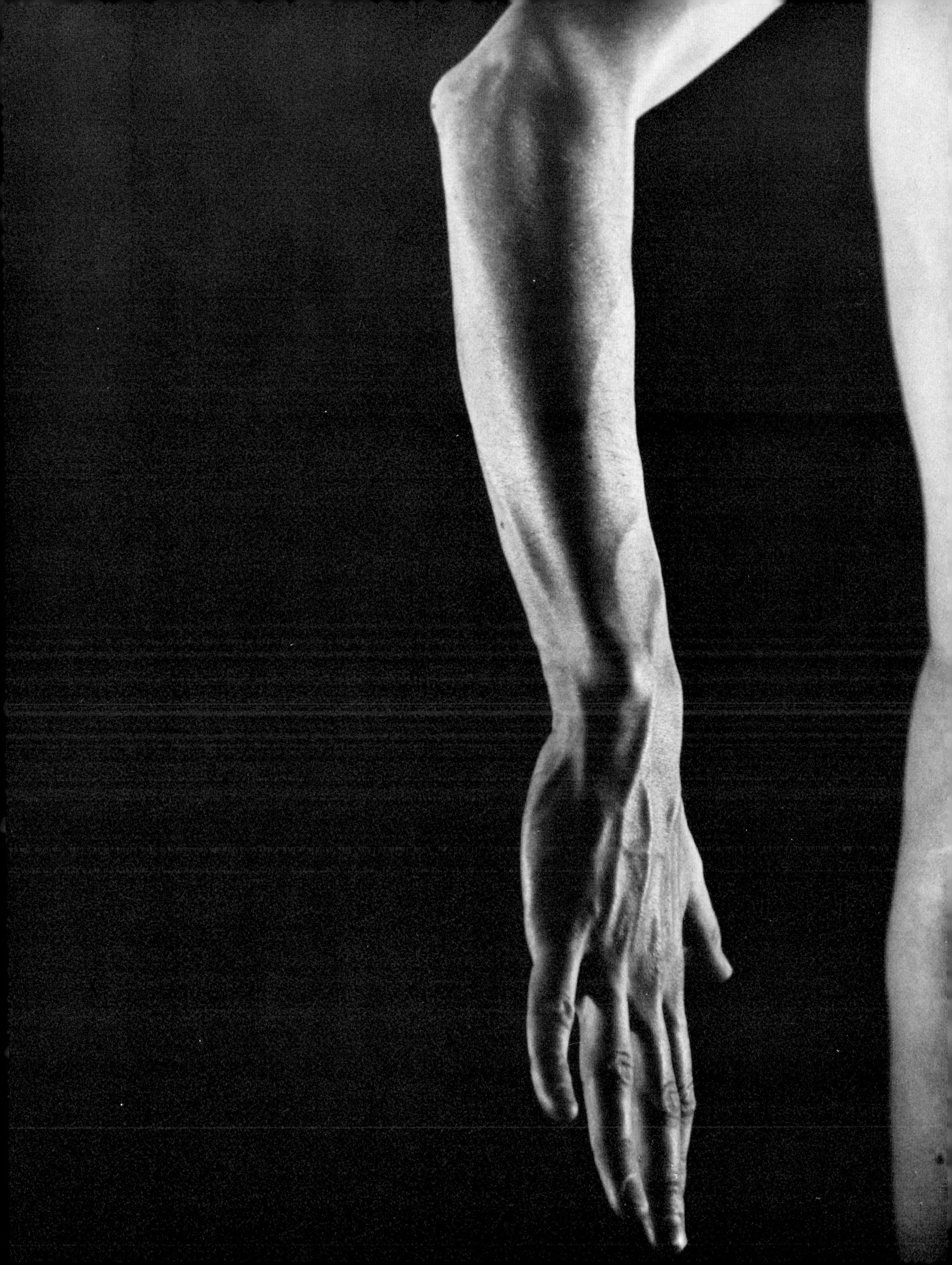

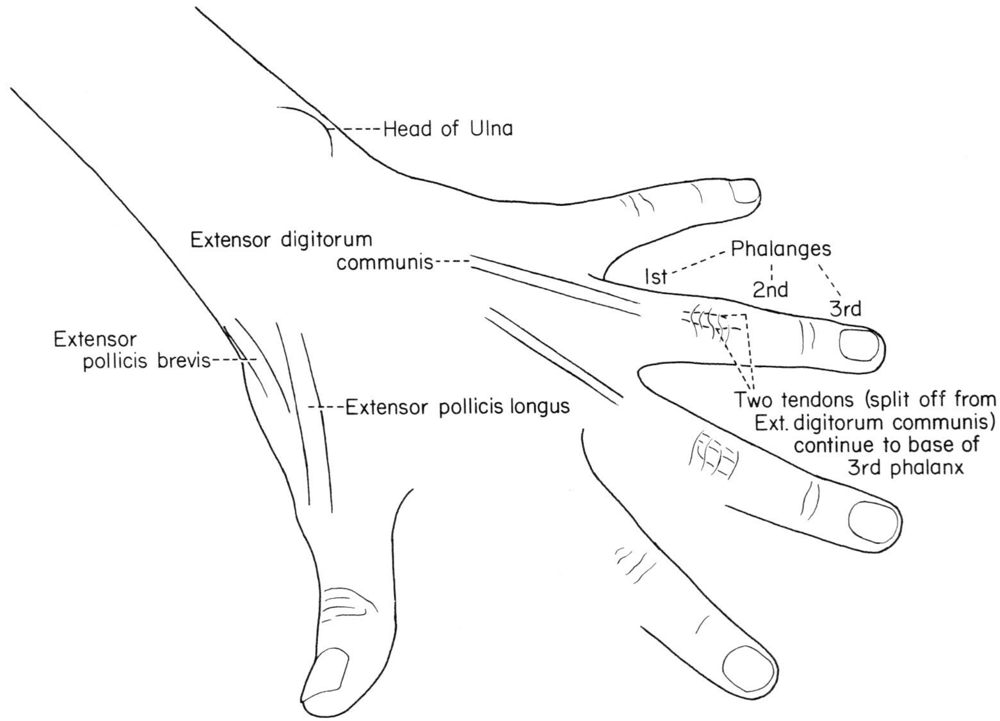

Figure 94 (♀ 10)

Strong abduction of the thumb brought out the tendons of the extensor pollicis muscles which proximally appear from underneath the ligamentum carpi dorsale, and which encompass a triangular-shaped depression. A slight depression on the finger joints (when they are extended fully) corresponds with the two split-off tendons of the extensor digitorum communis which continue together to the base of the third phalanx.

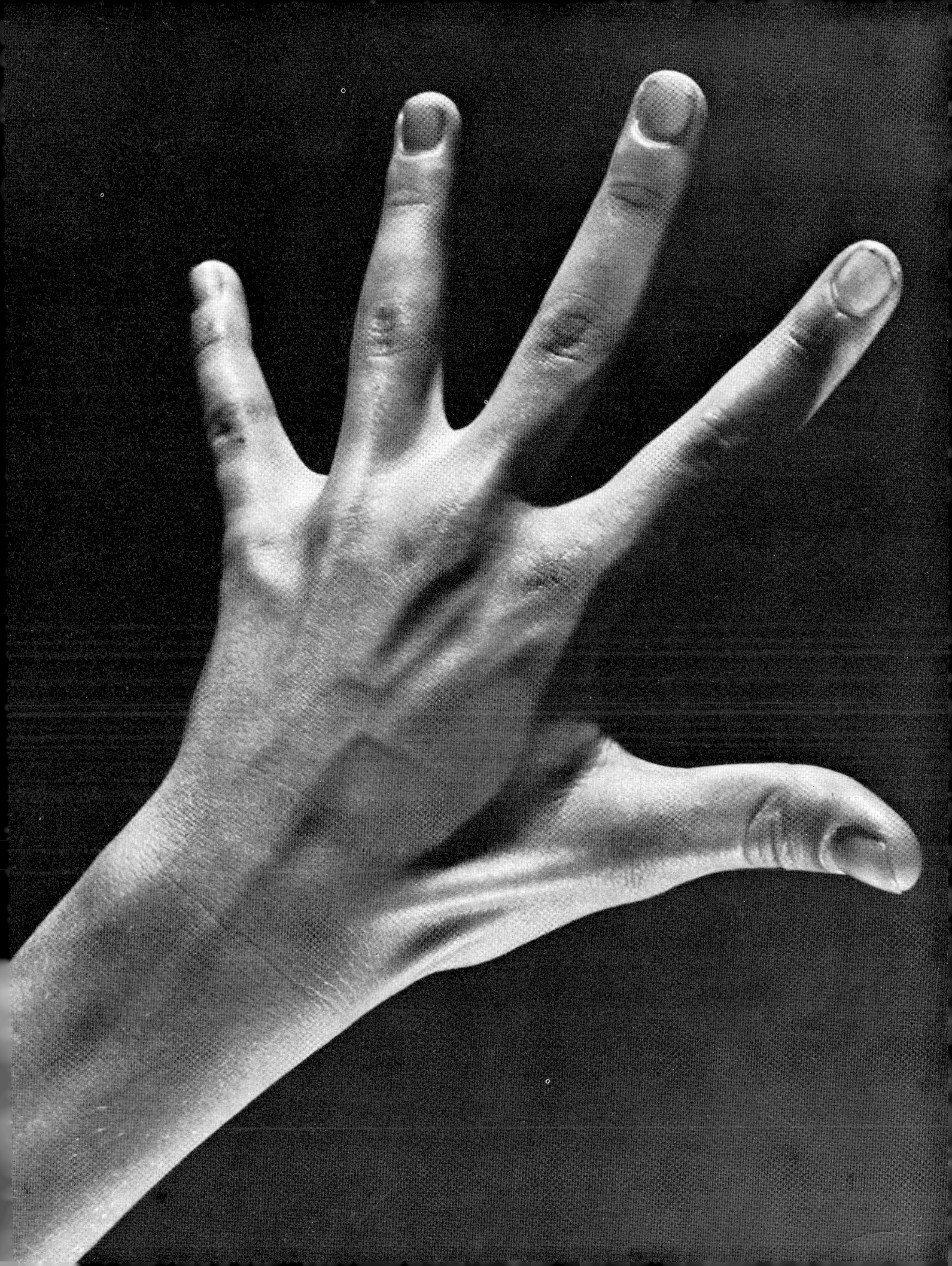

Figure 95 (♂ 23)

After working on an arm crank for five minutes and the subsequent occlusion (by air pressure cuff around the upper arm) of the circulation, the veins of the forearm and hand become clearly visible. Note the inter-connections between veins and the increase in size in the direction of the heart.

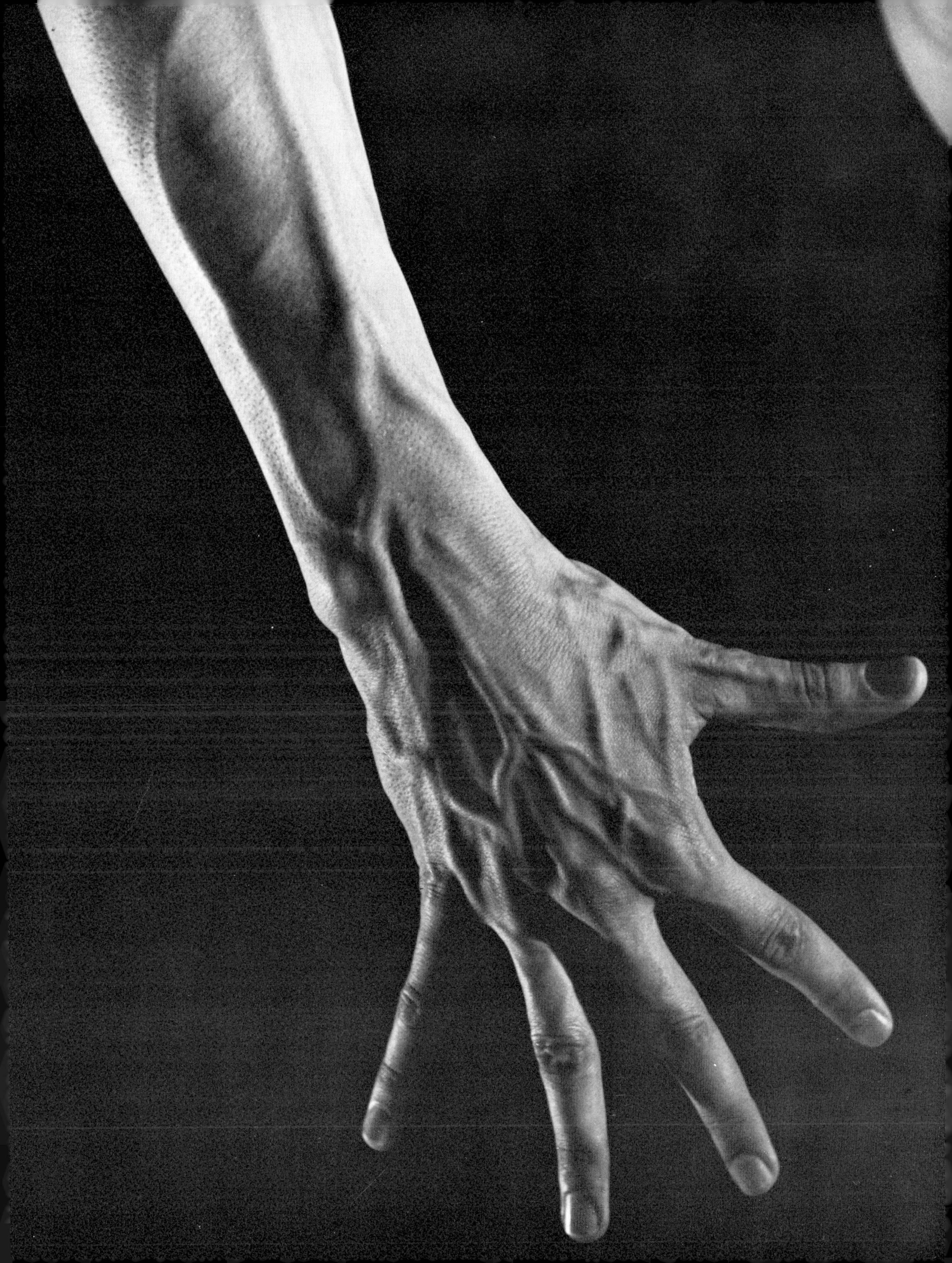

Figure 96 (♀ 52)

The skin has lost some resilience and has become loose. Contrast with Figures 93 and 97.

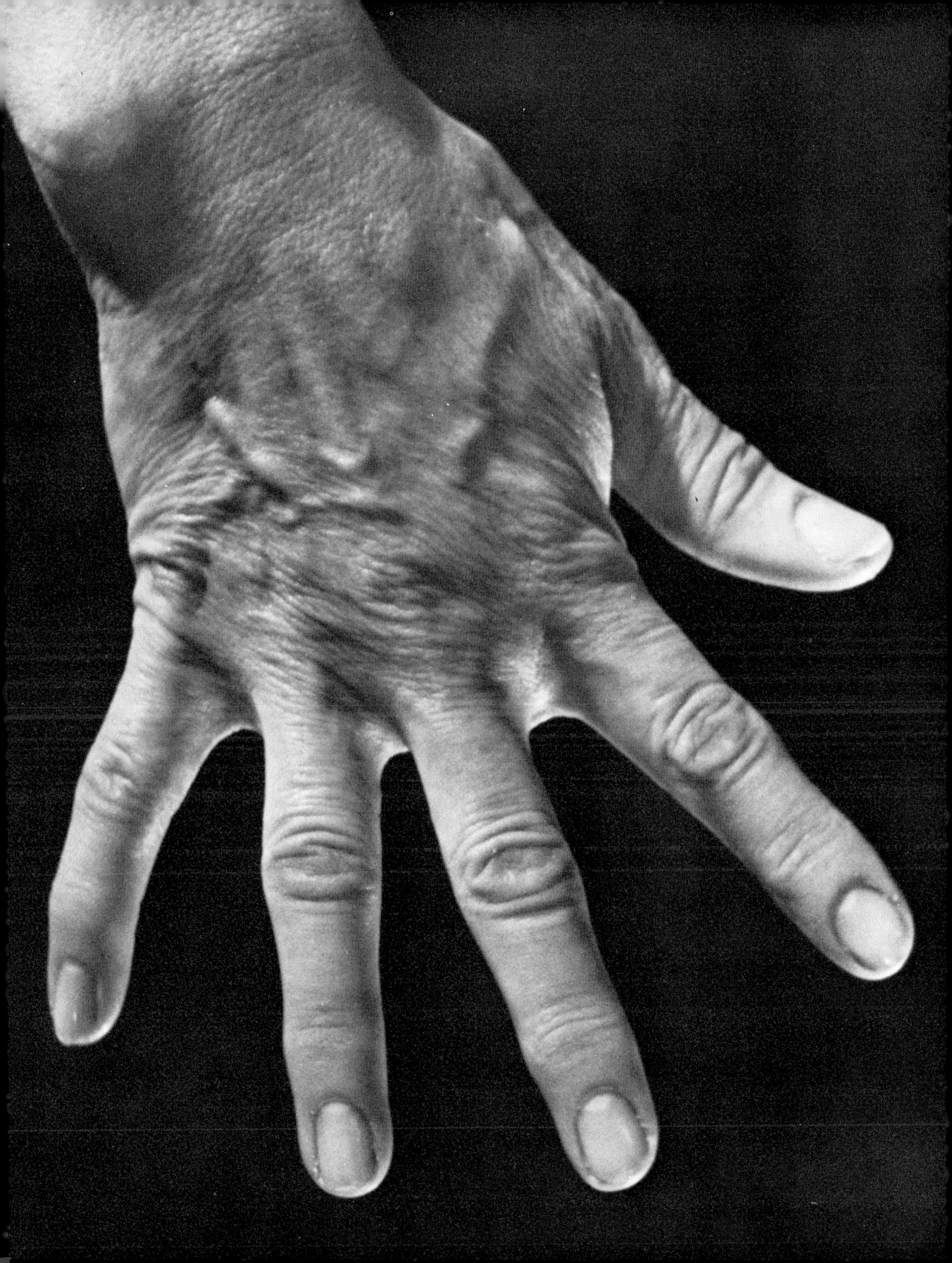

Figure 97 (♀ 17)

This subject was rather chubby. Her hand shows fatty deposits which obscure all but the faintest of tendon contours. The skin was firm and smooth.

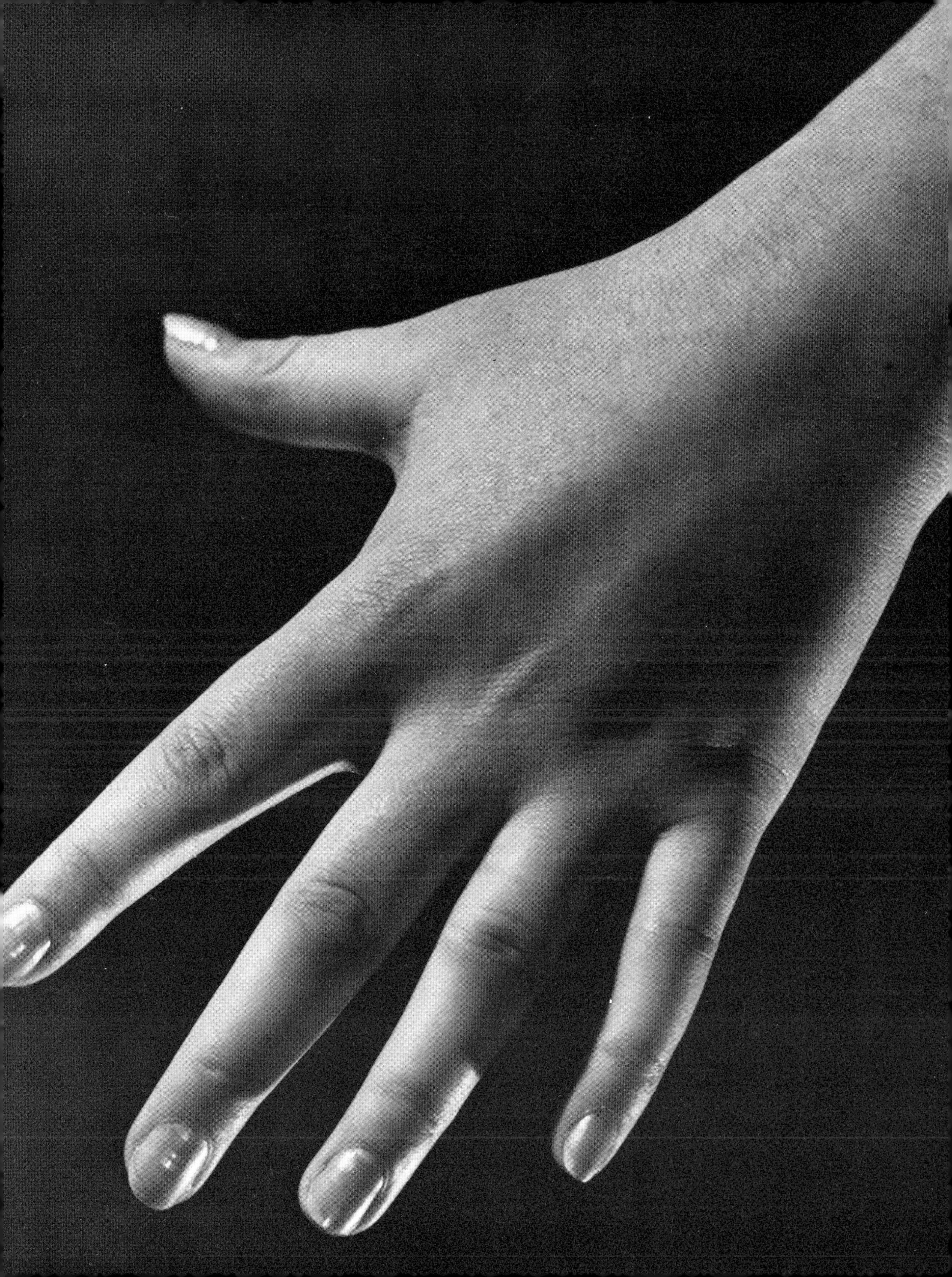

Figure 98 (♀ 5 days)

The hand of a baby does not show the underlying tissue contours but rather the sites of the fat pads. Creases on the skin indicate the site of the joints in these "sausages."

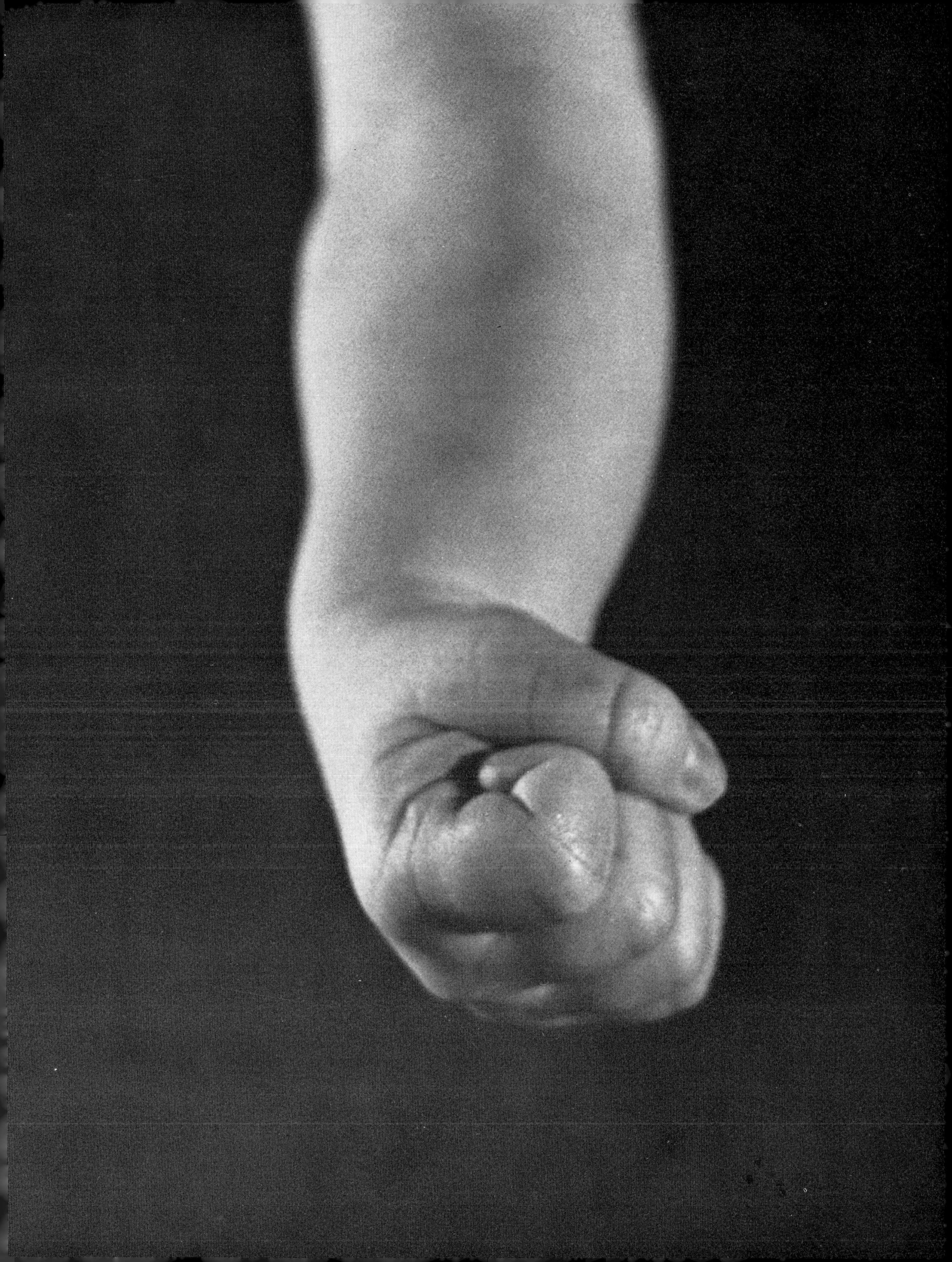

Figure 99 (♀ 5 days)

The skin is loosely fitted over the limbs, but is taut over the distended abdominal area. The remains of the umbilical cord are still present and will be lost in a few days. The legs are curved but will straighten out with growth. Proportionally, the head is much larger than in adults, while the legs and hips are smaller.

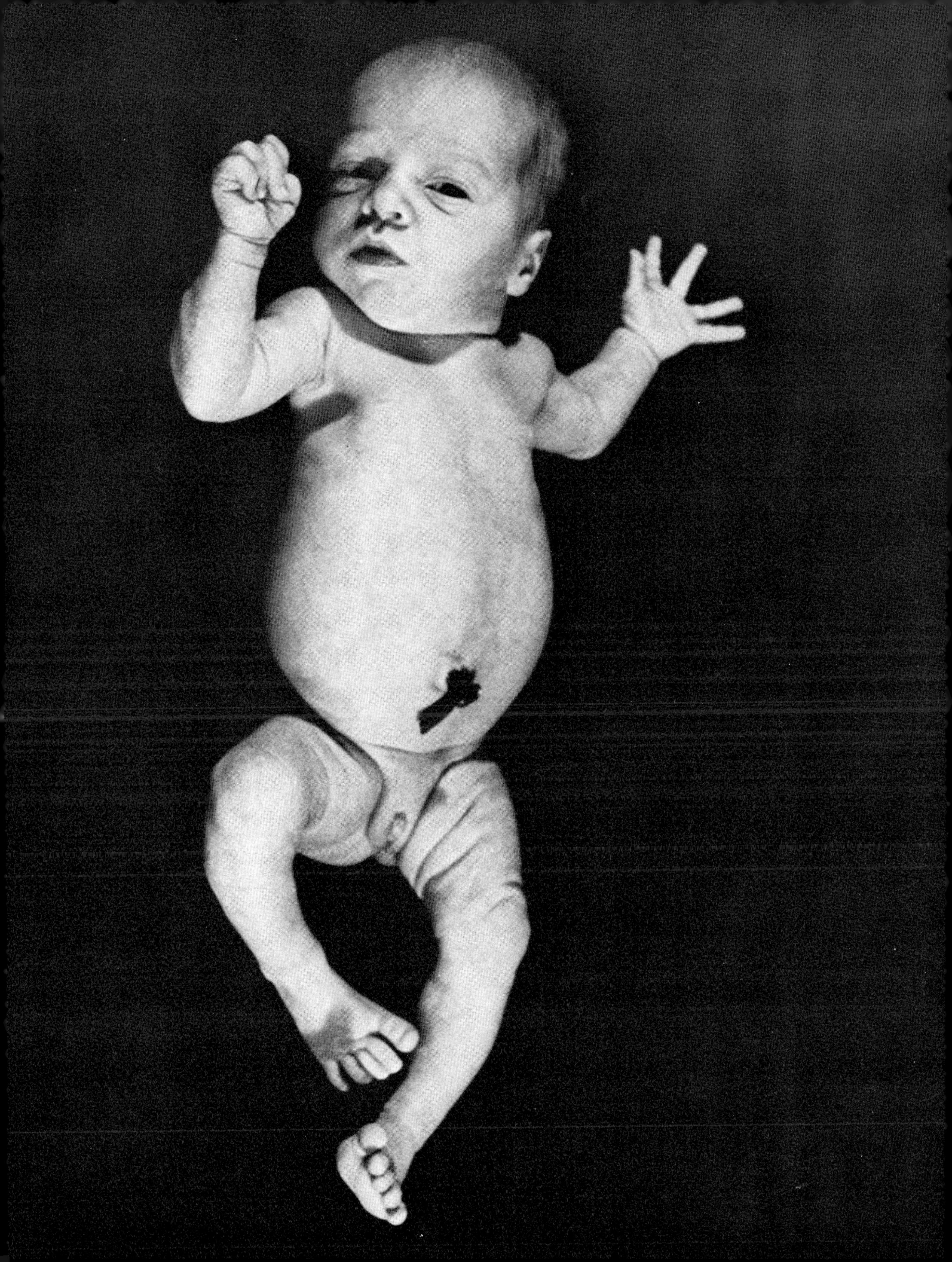

Figure 100 (♂ 1 month)

The skin has smoothed out over the limbs. Adipose tissue is being deposited. The distention of the abdomen may remain until the child loses his "babyish" look, somewhere around the age of six or seven years.

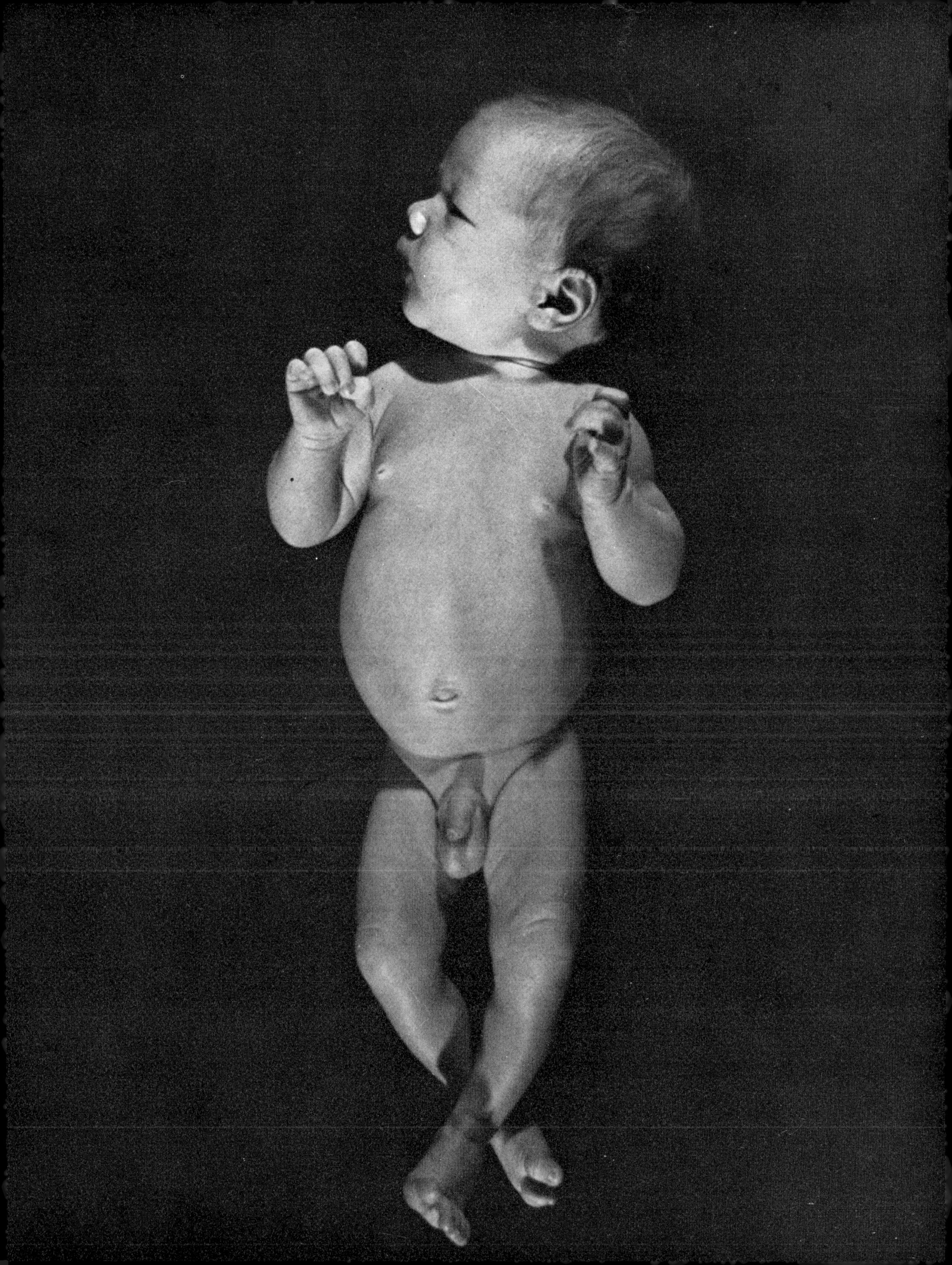

Figure 101 (♂ 1 year)

The limbs are straightening. Proportionally, the head and abdomen have become smaller.

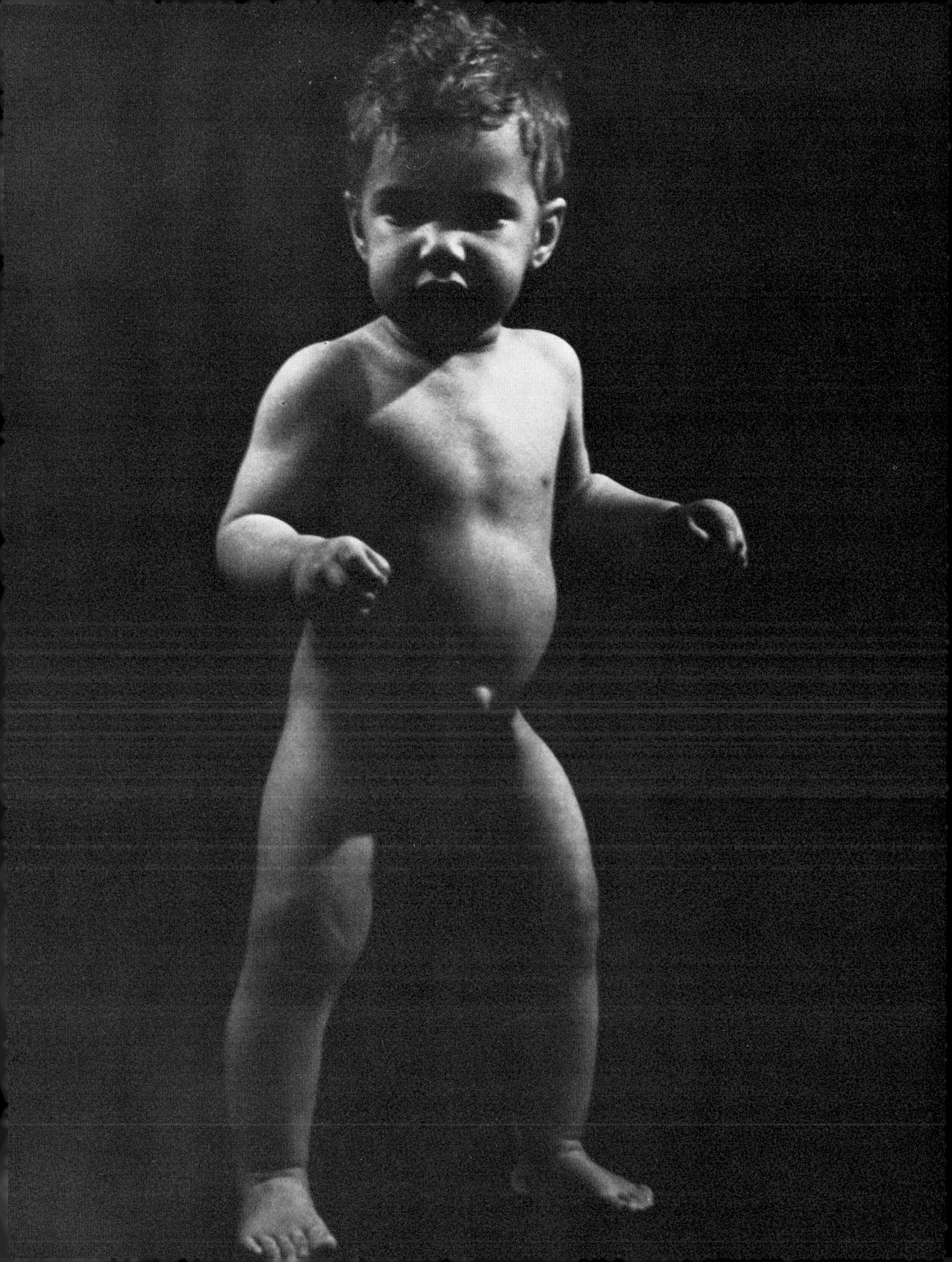

Figure 102 (♀ $2^1/_2$ years)

The bow legs have straightened. This little girl had even become slightly "knock-kneed" (genu valgum).

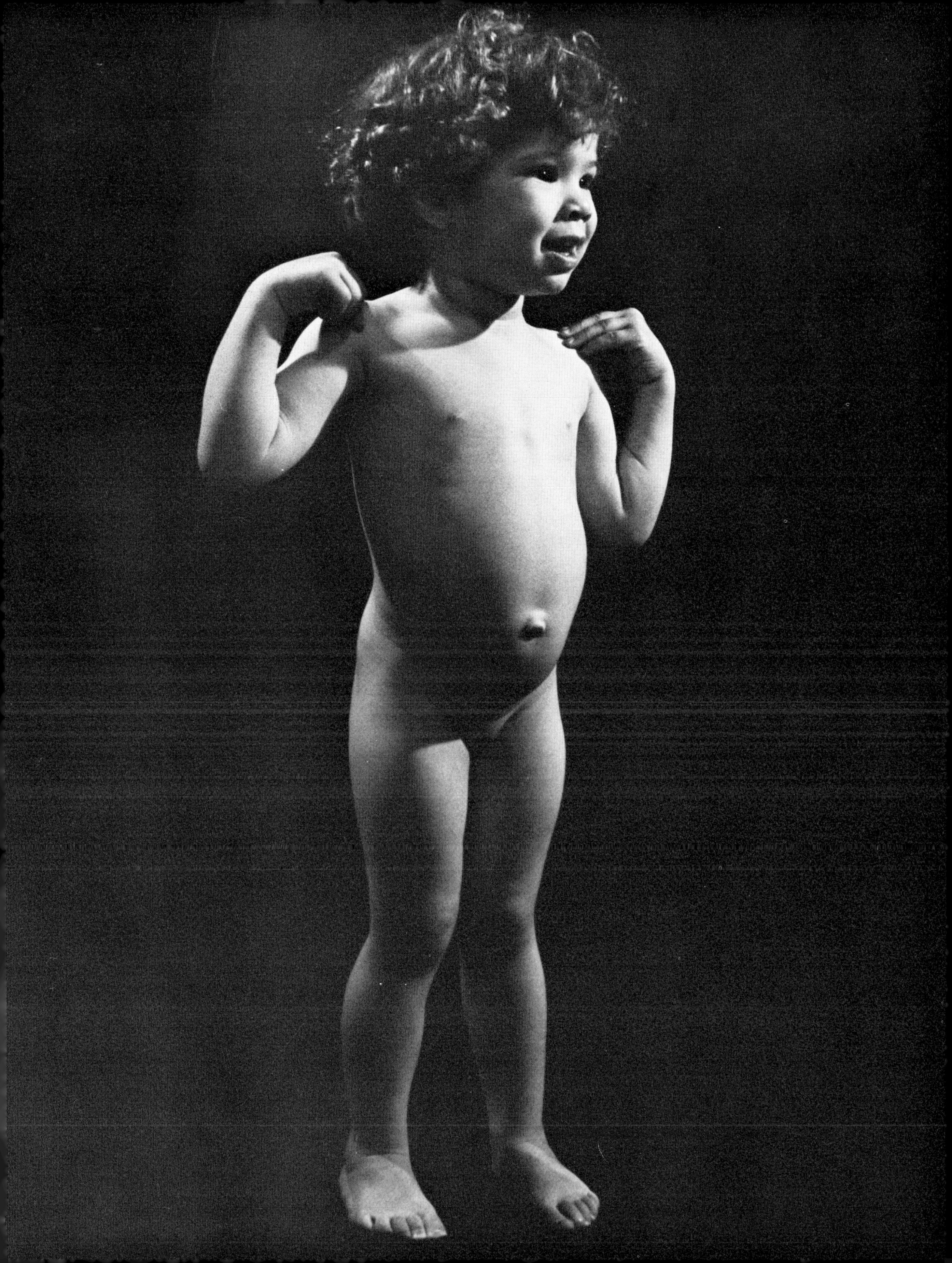

Figure 103 (♂ 5)

Musculature may be seen now, especially if the child is active and not fat. This child climbs a twenty-foot rope with ease, cycles, and bounces on a trampoline for hours at a stretch. Compare with the six-year-old girl in Figure 78.

Figure 104 (♀ 9)

The slightly overweight subject shows adipose tissue deposits mostly on the abdomen (mainly below the umbilicus and on the mons pubis) and the buttocks. Musculature and bony landmarks are (in this subject) difficult to discern.

Figure 105 (♂ 9)

A thin and muscular boy who is very active (all day) in climbing, running and cycling.

Figure 106 (♀ 11)

This thin girl had started to grow in limb length rapidly during the previous three months. She was quite athletic and strong. Musculature and bony landmarks are easily visible and palpated. The back view already shows definite feminine characteristics in the lower back and upper buttock region. The front view does not yet differ greatly from that seen in the boys in Figures 105 and 107.

Figure 107 (♂ 11)

This boy started a growth spurt six months before this photograph was taken. The thigh length increased noticeably. He was already active in many sports at this age.

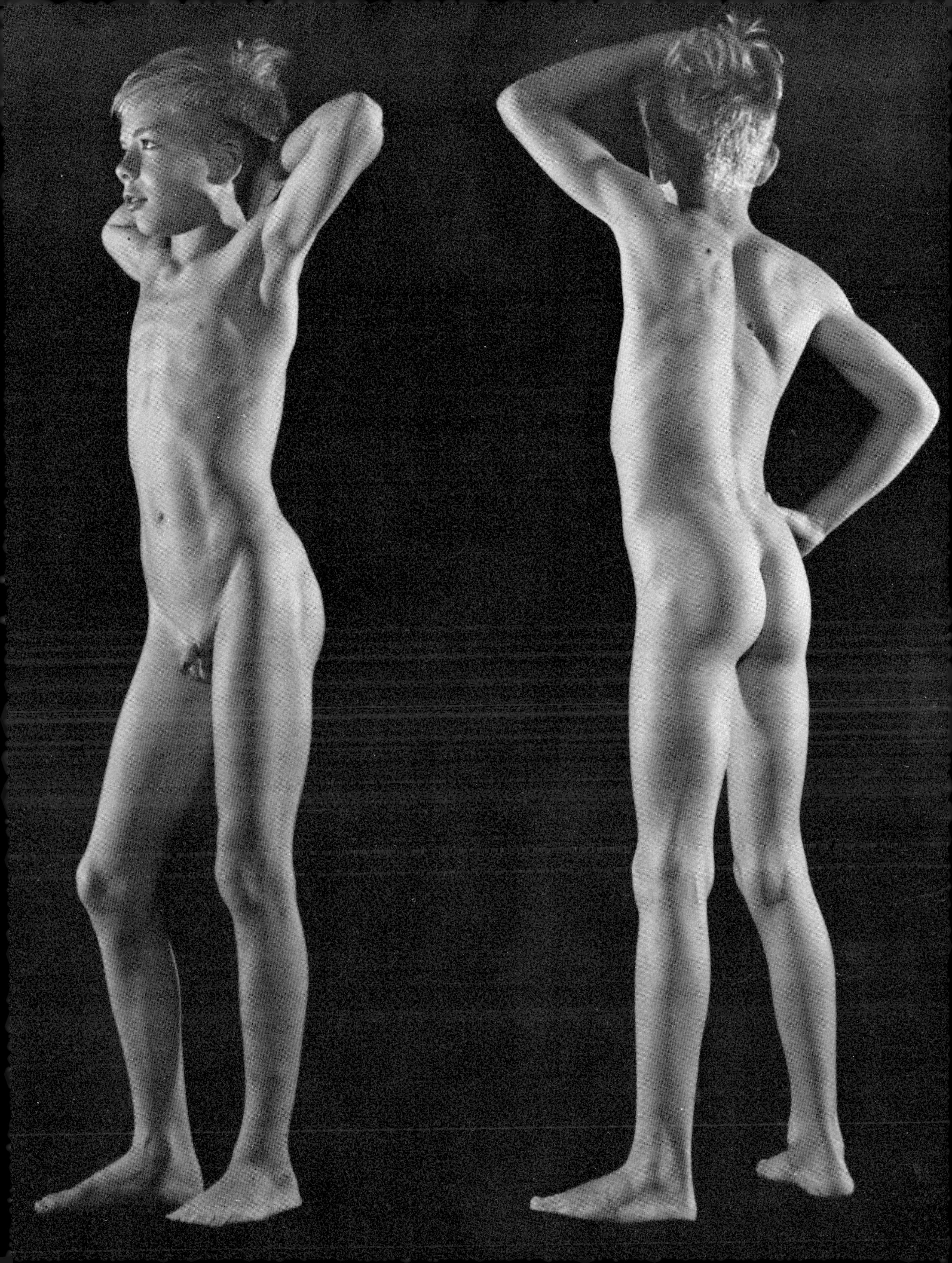

Figure 108 (♀ 13)

A late-maturing girl. Pubic hair was only visible for the two months previous to the taking of this picture; no menarche; the areolae have become pointed. General body contours had not changed for the previous two years.

Figure 109 (♀ 15)

The skeletal outline and the musculature on this extremely thin girl can be traced with ease. A good gymnast and sprinter. Late maturing.

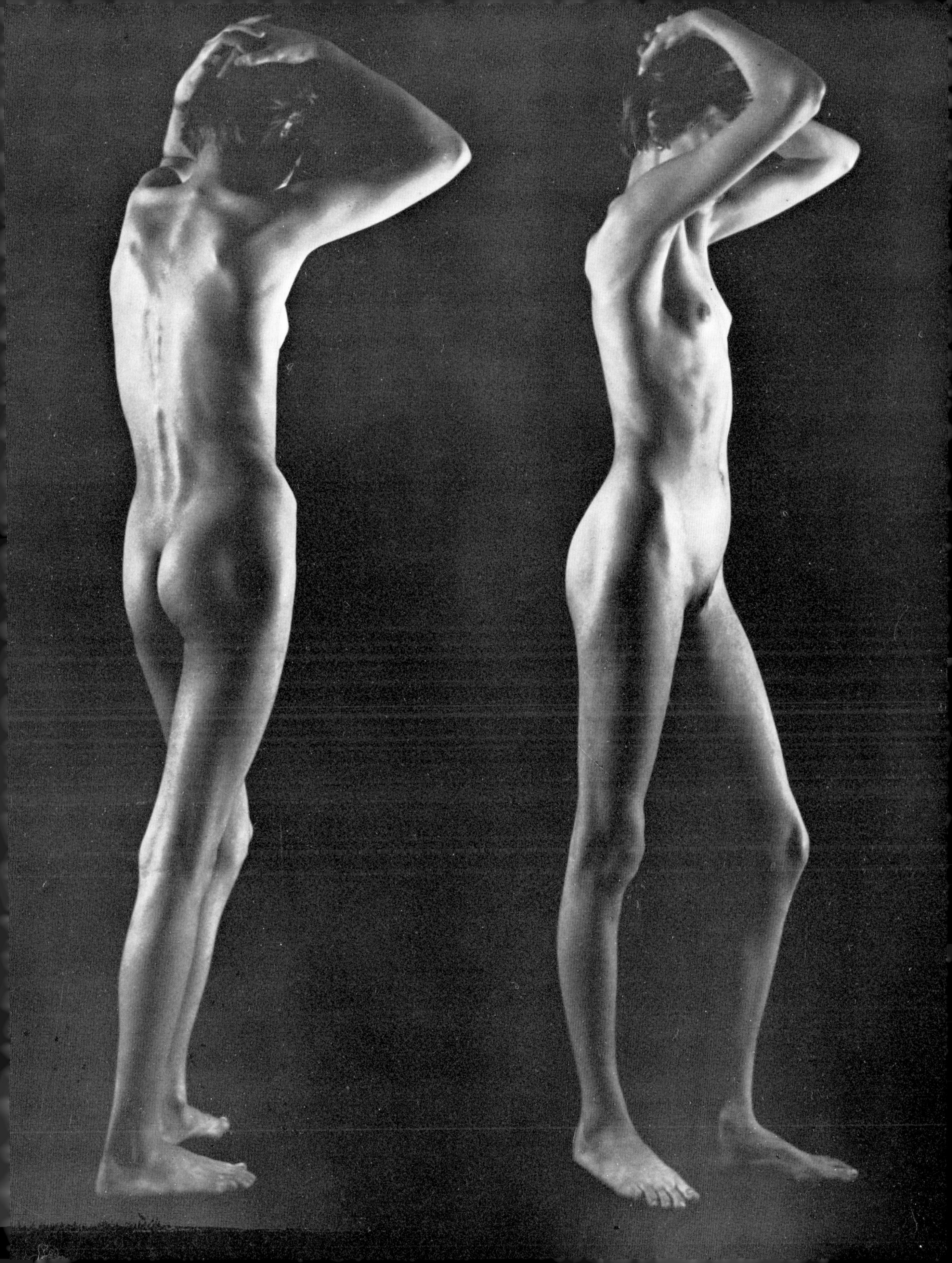

Figure 110 (♀ 16)

Normal maturity in a good gymnast. The hips have widened and adipose tissue has been deposited to give this girl a feminine (though not fully matured) look.

Figure 111 (♂ 16)

A tall subject who is a good distance runner and skater. The muscles are long and smooth.

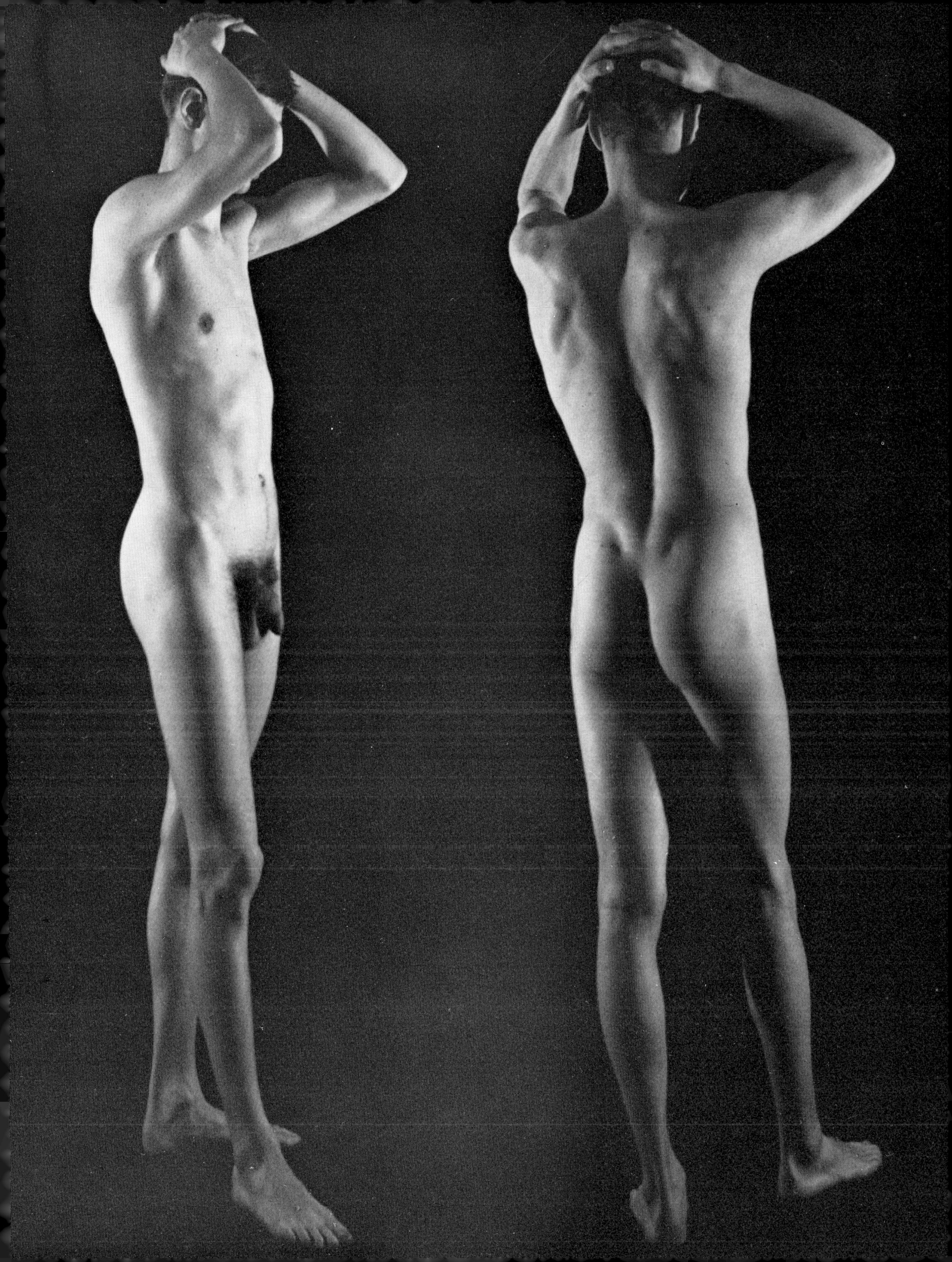

Figure 112 (♀ 19)

A nearly full-grown girl with fat padding on the hip and below the greater trochanter. The difference in the relative length of trunk to legs as compared to that in the average male has become clear-cut.

Figure 113 (♂ 19)

This subject was a good gymnast and strong in spite of his slightly chubby appearance. The abdominal wall appears to be bulging because of the abdominal contents. By contracting his muscles, however, the subject was able to flatten this surface and even "hollow" it.

Figure 114 (♀ 22, Japanese)

This young adult appears to be fully grown. Fat deposits hide a strong gymnast.

Figure 115 (♂ 22, Japanese)

A young adult with well-defined musculature. A championship-caliber sprint swimmer.

Figure 116 (♀ 24)

Young adult, overweight and rather physically inactive. Bony landmarks and musculature are hidden by fat deposits. Places that usually are well defined (axilla and inguinal groove) were filled and difficult to discern.

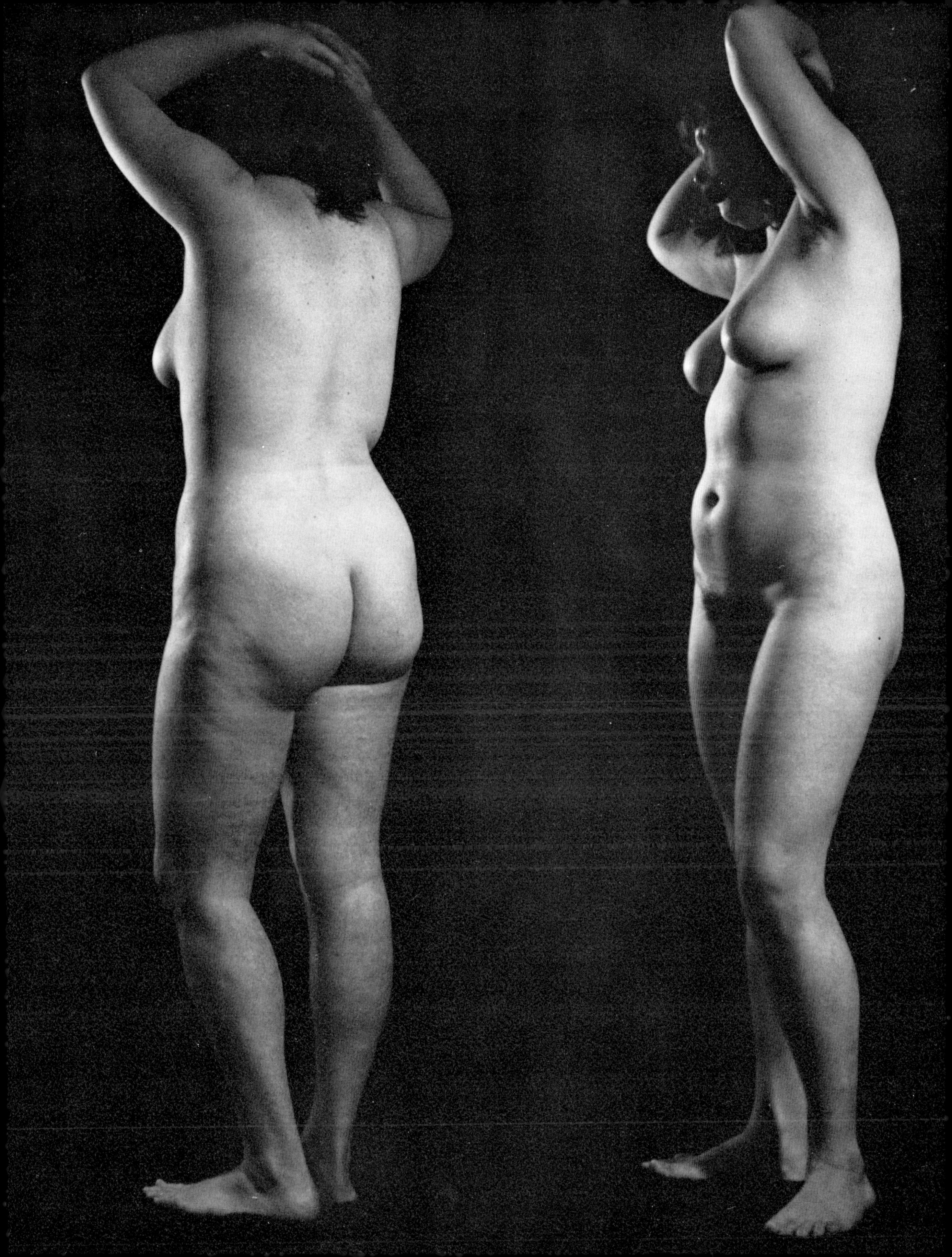

Figure 117 (♂ 24)

This tall, distance runner was extremely thin (in terms of fat deposits) but did not give this impression because of the well-developed musculature.

Figure 118 (♀ 30)

A mother of two children. The fat deposits seem to be concentrated around the hips and legs. The adipose bulge on the upper medial part of the thigh, formed in many adult females, is clearly visible. It is difficult to discern the patella. The dimples on the crista iliaca posterior superior show the tilt to the right with the compensatory lateral curvature of the spine to the left (compare Fig. 103).

Figure 119 (♀ 44)

This subject, a mother of two teenage children, was active in a gymnastics club twice a week for thirty-three years.

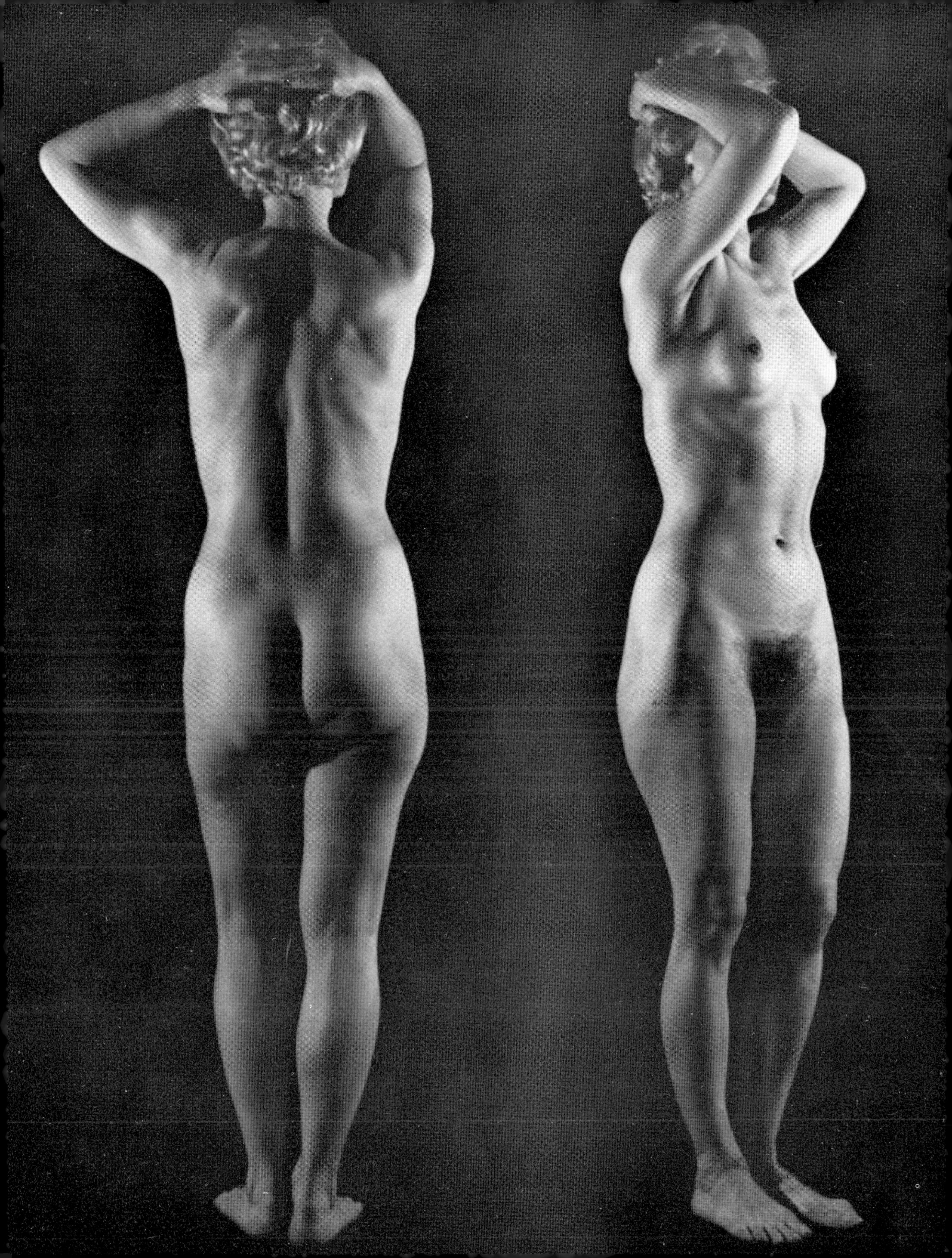

Figure 120 (♂ 48)

At middle age and still a good distance-runner (runs approximately twenty miles per week) this subject shows good muscle definition. He has participated in endurance exercises since he was a boy.

Figure 121 (♀ 50)

This subject has participated in tumbling and rhythmic exercise programs three times a week since she was ten years old. Although her abdominal muscles were strong (forty sit-ups with bent knees) she was unable to flatten her abdominal wall. Having an overweight problem she stated that losing more weight made her feel uncomfortable. Mother of two young adults.

Figure 122 (♀ 52)

A mother of four young adults, this subject has participated in gymnastics and many outdoor sports since she was thirty-eight years old. Her skin has lost much of its resiliency underneath the chin but is still rather smooth everywhere else.

Figure 123 (♂ 53)

This subject was a good hiker (four-day marches), though his muscle strength was not great. A deposit of adipose tissue is visible underneath his chin. His buttock muscles seem to have started to atrophy. The abdominal muscles were weak. He has participated in conditioning exercises irregularly for two years.

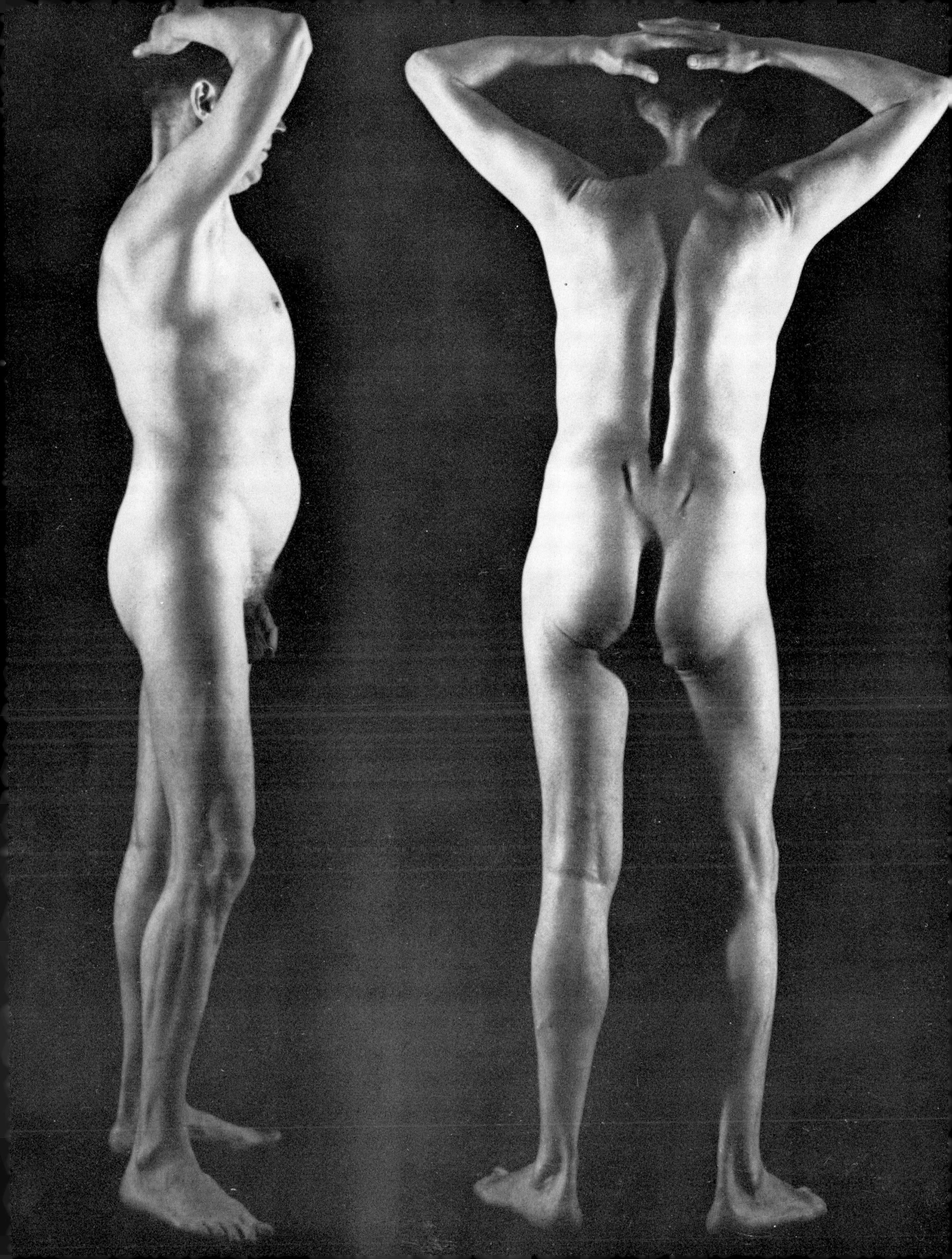

Figure 124 (♀ 62)

Despite her age and weight this large woman was able to run in a very light-footed manner. She has participated in gymnastics and rhythmic conditioning exercises since the age of thirty years. Adipose tissue seems to keep her skin smooth everywhere, except where it shows up clearly laterally of the breasts and on the lower abdomen.

Figure 125 (♂ 72)

The skin has lost much of its resiliency and hangs loosely underneath the chin and over the triceps. The abdominal wall was pushed out by the abdominal contents, but was not weak in musculature (thirty sit-ups with bent knees, with ease). The muscles in general were strong (thirty push-ups), the weakest being the knee and hip extensors. The gluteal fold is formed by loose skin rather than by a clear division of gluteus maximus and hamstring muscles. Varicose veins on the lower legs were present at the age of forty years but never caused any discomfort. This subject has participated in conditioning exercises since the age of twelve years; he gave up bicycle-racing at forty.

Figure 126 (♀ 75)

Many muscle groups have atrophied considerably in the present subject. Despite this she was able to participate in many tumbling and rhythmic gymnastics activities (which she has done five times per week since the age of eleven years), perhaps because of the predominant role played by skill rather than muscle strength. The abdominal wall was weak and showed distinct areas of loss of skin resiliency. The buttocks have flattened, and, together with the adipose tissue above them, give the impression of a very long, flat-backed trunk and short legs. The upper part of the spine has increased its curvature (kyphosis).

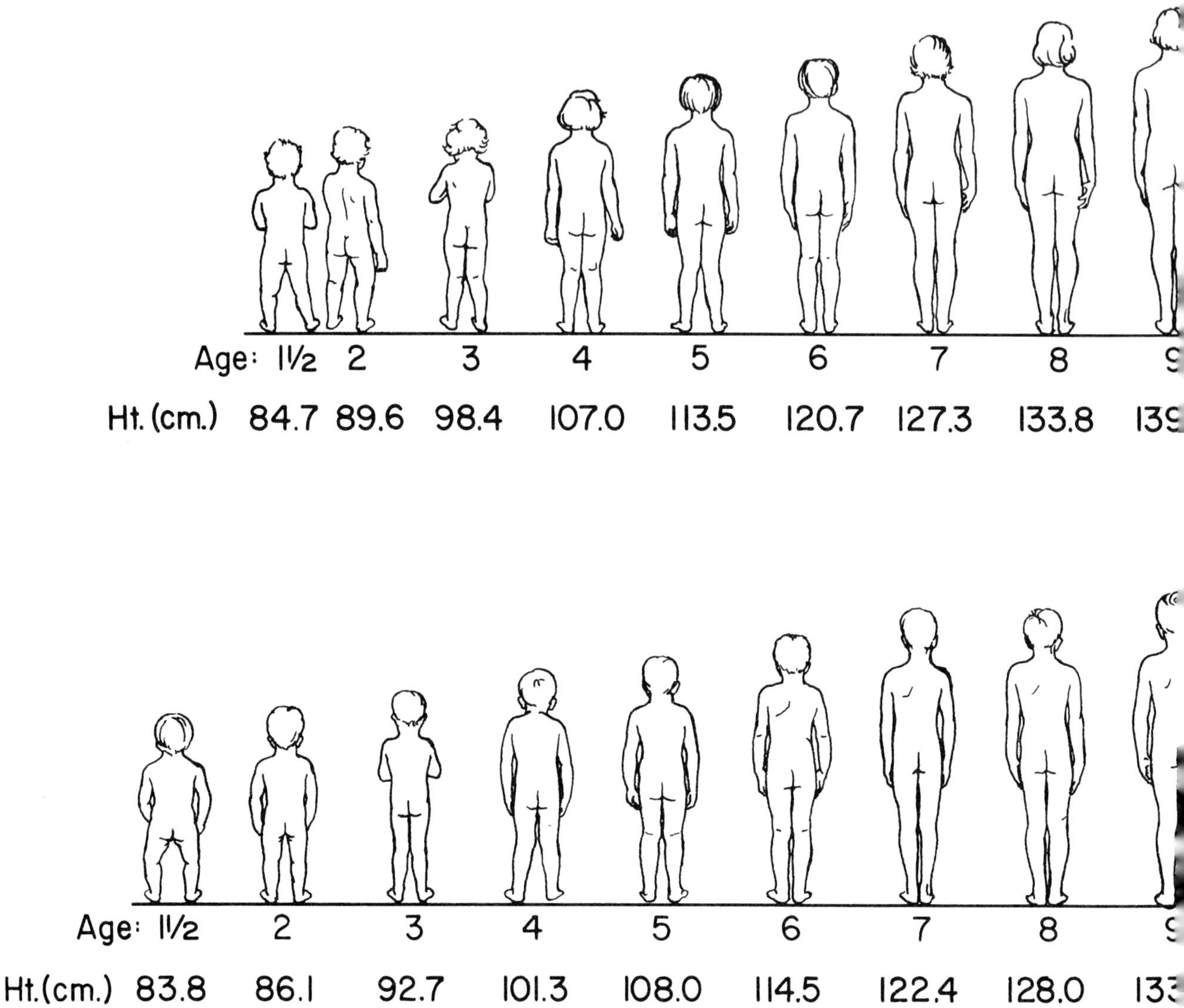

Figure 127

Outline drawing of boy and girl illustrating changes in proportions and stance (eighteen months to eighteen years). (From Bayer, L. M., and Bayley, Nancy: Growth Diagnosis. University of Chicago Press, Chicago, 1959, pp. 24 and 25. Reproduced with permission of the authors and publisher.)

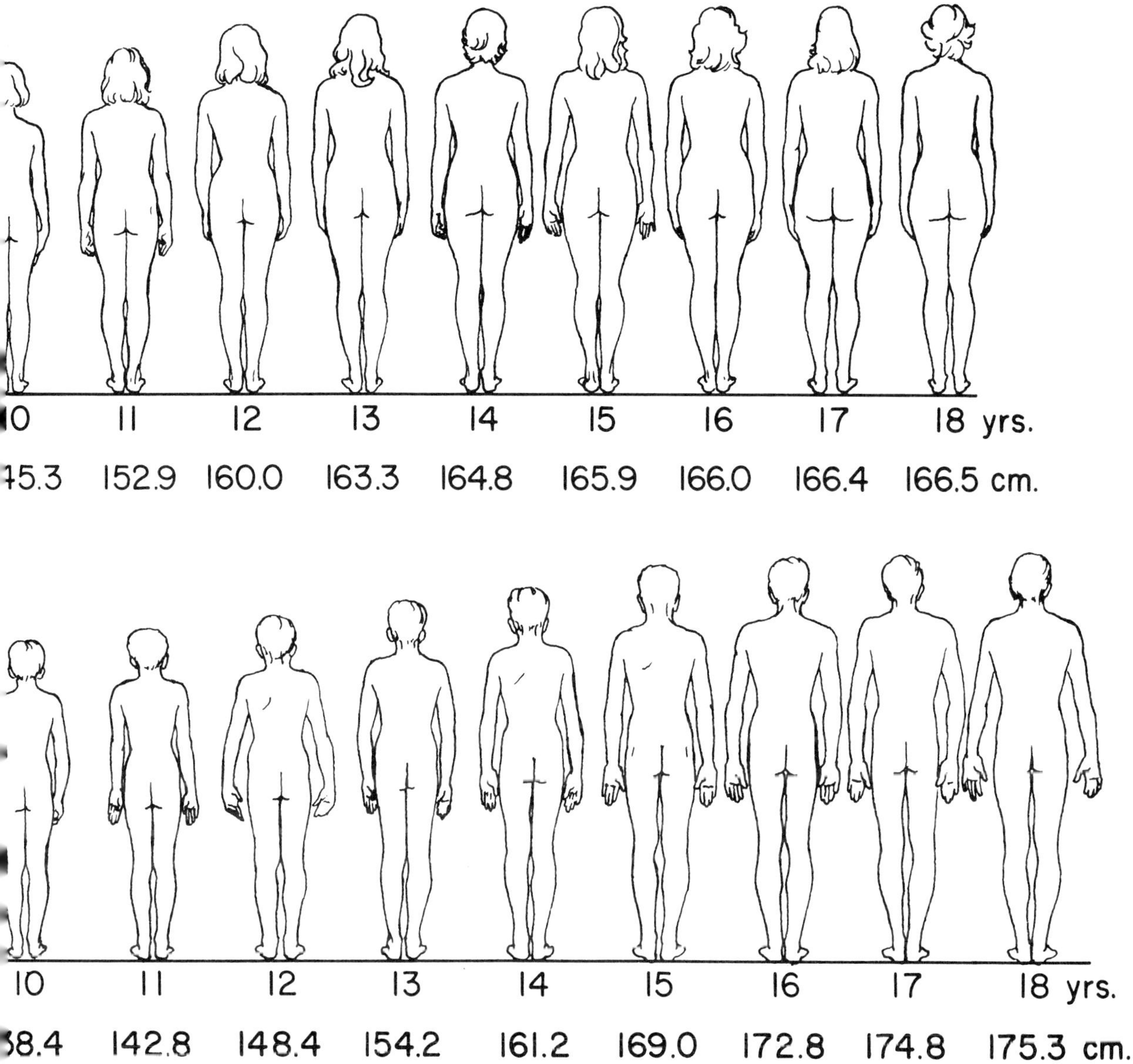
10 11 12 13 14 15 16 17 18 yrs.
45.3 152.9 160.0 163.3 164.8 165.9 166.0 166.4 166.5 cm.
10 11 12 13 14 15 16 17 18 yrs.
38.4 142.8 148.4 154.2 161.2 169.0 172.8 174.8 175.3 cm.

Index

Abdomen, 84, 85, 92, 93, 102, 103
  adipose tissue deposits on, of adolescent female, 88, 89
  chest and, 68, 69
  in six-month pregnant woman, 90, 91
  muscles of, 82, 83
  post-pregnancy stretch marks on, 104, 105
  ribs and, 72, 73
  thigh and, venous pattern of, 108, 109
  wall of
    flattening of, 98, 99, 100, 101, 106, 107
    ribcage and, 96, 97
    "washboard" effect of, 94, 95
Abductor digiti V, 170, 182
Abductor pollicis brevis, 170, 182
Abductors, hip, 124, 125
Achilles tendon, 52, 128, 138, 160
Acromion, 4, 16, 17, 22, 24, 38, 56
Adam's apple, 16
Adductor hallucis, 166
Adductor longus, 110
Adductors, 110, 112, 114
Adipose tissue
  deposits of, on abdomen of adolescent female, 88, 89
  variation in, 60, 61, 62, 63
Adolescent. See *Teen-ager*
Adult female (full figure of)
  age 22 years, 229
  age 24 years, 233
  age 30 years, 237
  age 44 years, 239
  age 50 years, 243
  age 52 years, 245
  age 62 years, 249
  age 75 years, 253
Adult male (full figure of)
  age 22 years, 231
  age 24 years, 235
  age 48 years, 241
  age 53 years, 247
  age 72 years, 251
Anconeus, 178, 180
Areola mammae, 74, 75, 78, 79
Arm, 172, 173, 174, 175, 184, 185
  hand and, 176, 177, 178, 179, 180, 181, 186, 187
  neck, shoulder, chest and, 170, 171
Aponeurosis plantaris, 166

Back, 50, 51, 56, 57
  shoulders and, 36, 37, 38, 39
Basilic vein, 182, 183
Biceps, 26, 36, 170, 172, 176, 178, 182
  action of, on forearm, schematic representation of, 2
Biceps femoris, 66, 122, 132, 136, 138, 140, 144
  tendon(s) of, 58, 59, 116, 128, 146, 147
Body
  bony landmarks of, 1, 4
    identification of, 1, 3, 4
Body (full figure of)
  female adult
    age 22 years, 229
    age 24 years, 233
    age 30 years, 237
    age 44 years, 239
    age 50 years, 243
    age 52 years, 245
    age 62 years, 249
    age 75 years, 253
  female child
    age $2^1/_2$ years, 205
    age 9 years, 209

Body (full figure of) *(continued)*
female child *(continued)*
age 11 years, 213
female infant, age 5 days, 199
female teen-ager
age 13 years, 217
age 15 years, 219
age 16 years, 221
age 19 years, 225
in flexion, 58, 59
in prone position, 52, 53, 54, 55
major muscle groups of, role of, 3, 4-13
male adult
age 22 years, 231
age 24 years, 235
age 48 years, 241
age 53 years, 247
age 72 years, 251
male child
age 1 year, 203
age 5 years, 207
age 9 years, 211
age 11 years, 215
male infant, age 1 month, 201
male teen-ager
age 16 years, 223
age 19 years, 227
motion of
complexity of, 1
mechanics of, 1
muscles of, identification of, 3
types of levers in, 1
Body proportions and stance, changes in, from 18 months to 18 years, 254, 255
Bony landmarks of human body, 1, 4
identification of, 1, 3, 4
Brachialis, 170, 176
Brachioradialis, 176, 182
Breasts, 24, 40, 41, 72, 73, 74, 75, 80, 81, 104, 105
lactating, 76, 77
virginal, 78, 79

Calcaneus, 128, 144
Cartilage
cricoid, 16
thyroid, 16
Cervical spine, 38
7th, 42, 44
Cheekbones, 14, 15
Chest
abdomen and, 68, 69
arm, neck, shoulder and, 170, 171
head, shoulders and, 30, 31, 32, 33
neck, shoulders and, 24, 25
shoulders and, 40, 41
Child (full figure of)
female
age $2^1/_2$ years, 205
age 9 years, 209
age 11 years, 213
male
age 1 year, 203
age 5 years, 207
age 9 years, 211
age 11 years, 215
Clavicle, 14, 15, 16, 20, 22, 24, 94, 170
attachment of, 14
Coccyx, 42, 43
Condyle
lateral, anterior aspect of, 110
medial, 110, 116
Coracobrachialis, 24, 26, 34
Coracoid processes, 20
Costal margin, 68
Cricoid cartilage, 16
Crista iliaca, 118
anterior superior, 4, 68, 126
posterior superior, 56

Deltoid, 22, 24, 26, 36, 52, 170, 174, 176
Digitation of serratus anterior and obliquus externus, 84

Elbow, 184, 185
Epicondyle
lateral
of femur, 4
of humerus, 4, 22, 174, 180
medial
of femur, 4
of humerus, 22
Extensor carpi radialis brevis, 176, 178
Extensor carpi radialis longus, 35, 174, 176, 178
Extensor carpi ulnaris, 36, 172, 176, 178, 180
Extensor digitorum brevis, 164
Extensor digitorum communis, 172, 176, 178, 180, 188

Extensor digitorum longus, 160, 161, 162, 164
tendons of, 144
Extensor hallucis longus, 144, 164
Extensor pollicis brevis, 174, 188
Extensor pollicis longus, 174, 188

Fascia axillaris, 26
Fascia pectoralis, 18
Fat deposits, unevenly distributed, 60, 61, 62, 63
Female (full figure of)
adult
age 22 years, 229
age 24 years, 233
age 30 years, 237
age 44 years, 239
age 50 years, 243
age 52 years, 245
age 62 years, 249
age 75 years, 253
child
age 2½ years, 205
age 9 years, 209
age 11 years, 213
infant, age 5 days, 199
teen-ager
age 13 years, 217
age 15 years, 219
age 16 years, 221
age 19 years, 225
Femur
lateral epicondyle of, 4
medial epicondyle of, 4
Fibula, head of, 4, 122
Flexion, curvature of spine in, 58, 59
Flexor carpi radialis, 170, 174, 182
Flexor carpi ulnaris, 170, 172, 178, 180
Flexor digiti V, 170, 182
Flexor digitorum brevis, 166
Flexor digitorum sublimis, 182
Flexor policis brevis, 170
Foot, 160, 161, 162, 163, 164, 165, 166, 167
of infant, 168, 169
Force of lever system, 1
Forearm
action of biceps on, schematic representation of, 2
action of triceps on, schematic representation of, 2
hand and, 190, 191
Fossa, greater supraclavicular, 16, 17
Fulcrum of lever system, 1
Full figure. See *Body (full figure of)*

Gastrocnemius, 52, 122, 128, 132, 138, 140, 144, 158
Genu recurvatum, 156, 157
Genu varum, 156, 157
Glans penis, 116, 117
Gluteal fold, 52, 56, 132, 136, 138
Gluteus maximus, 52, 110, 112, 114, 132, 134, 135, 136
Gluteus medius, 52, 66, 84, 85, 110, 122, 132, 136
Gracilis, 140
Groin, 68
in female, 112, 113, 114, 115
in male, 110, 111, 116, 117

Hamstrings, 52, 128, 129, 138, 139
Hand, 188, 189, 192, 193, 194, 195
arm and, 176, 177, 178, 179, 180, 181, 186, 187
forearm and, 190, 191
of infant, 196, 197
Head
maintenance of equilibrium of, 14
neck and, 14, 15
shoulders and, 20, 21
chest and, 30, 31, 32, 33
Hip, 124, 125, 126, 127
legs and, 132, 133, 134, 135, 136, 137, 138, 139, 142, 143
Hip abductors, 124, 125
Hip joint, 126, 127
flexion in, 130, 131
Human body. See *Body*
Humerus
medial epicondyle of, 22
lateral epicondyle of, 4, 22, 174, 180

Iliofemoral ligament, 66
Iliotibial tract, 66, 116, 122, 128, 136, 144
Illium, posterior superior spines of, 48, 50
Infant
foot of, 168, 169
full figure of
female, age 5 days, 199
male, age 1 month, 201
hand of, 196, 197

Infraspinatus, 34, 36, 174
Inguinal groove
  in female, 112, 113, 114, 115
  in male, 110, 111, 116, 117
Inguinal ligament, 106, 107
Inscriptio tendinea of rectus abdominis, 82, 83, 94
Interosseus dorsalis, 170, 176

Jaw, 14, 15
Joint
  hip, flexion in, 130, 131
  shoulder, 28, 29
    adduction in, 26, 27
    static horizontal, flexion-adduction in, 32, 33
  sternoclavicular, 16

Knee(s), 128, 129, 152, 153, 156, 157
  passive flexion in, 154, 155

Labium majus, 112, 113, 114, 115
Labium minus, 112, 113, 114, 115
Lactating breasts, 76, 77
Landmarks, bony, of human body, 1, 4
  identification of, 1, 3, 4
Lateral epicondyle, 4, 22, 174, 180
Lateral malleolus, 4, 52, 144, 158, 164
Latissimus dorsi, 24, 26, 34, 68, 84, 174
  anterior edge of, 68
Leg(s), 140, 141, 144, 145, 146, 147, 148, 149, 150, 151, 152, 153, 158, 159
  hips and, 132, 133, 134, 135, 136, 137, 138, 139, 142, 143
  thighs and, 128, 129
  trunk and, 120, 121, 122, 123, 130, 131
Lever(s)
  skeleton as structure of, 1
  system of, constituents of, 1
  types of, in human body, 1
Ligament
  iliofemoral, 66
  inguinal, 106, 107
  patellar, 144
Ligamentum carpi dorsale, 176
Ligamentum cruciatum cruris, 164
Limbs, trunk and, 118, 119
Linea alba, 68, 82, 83, 84, 98, 99, 118
Linea semilunaris, 68, 84

Lombard's paradox, 128
Lumbar vertebrae, 42, 43

Male (full figure of)
  adult
    age 22 years, 231
    age 24 years, 235
    age 48 years, 241
    age 53 years, 247
    age 72 years, 251
  child
    age 1 year, 203
    age 5 years, 207
    age 9 years, 211
    age 11 years, 215
  infant, age 1 month, 201
  teen-ager
    age 16 years, 223
    age 19 years, 227
Malleolus lateralis, 4, 52, 144, 158, 164
Malleolus medialis, 4, 158, 162, 164, 166
Mandible, 14, 15
Manubrium, 20
Medial border of scapula, 4, 36
Medial epicondyle, 22, 170
Medial popliteal nerve, 128
Metatarsal I, 166
Motion of human body
  complexity of, 1
  coordinated, 1
  mechanics of, 1
Mouth, neck and, 18, 19
Movement. See *Motion*
Muscle(s). See also under specific name, e.g., *Trapezius*
  identification of, 3
Muscle groups, major, of human body, role of, 3, 4-13

Neck
  head and, 14, 15
  mouth and, 18, 19
  shoulders and, 16, 17, 22, 23
    chest and, 24, 25
      arm and, 170, 171
  trunk and, 42, 43
Nerve, medial popliteal, 128

Obliquus externus, 68, 82, 83, 94, 95, 110, 116

Obliquus externus *(continued)*
serratus anterior and, digitation of, 84, 94
Obliquus internus, 68, 82, 83
Olecranon, 4, 22, 36, 174, 176, 186, 187
Orbicularis oris, 18
Os ischii, 114
Os pubis, 114

Palmaris longus, 170, 182
Patella, 80, 81, 116, 122, 128
Patellar ligament, 144
Patellar tendon, 128
Pectoralis major, 20, 22, 24, 26, 30, 32, 34, 68, 84, 94, 170
Pelvis, 80, 81, 116
Penis, 110, 111, 116, 117
glans, 116, 117
Peroneus brevis, 140, 144, 160, 162
Peroneus longus, 138, 144, 158, 160, 162
Peroneus III, 160, 162, 164
Phalanges, 188
Platysma, 18
Popliteal nerve, medial, 128
Popliteal space, 132, 138
Pregnancy, abdomen in, 90, 91
Puberty, onset of, in female, fattening process in, 88, 89
Pubic arch, 4, 66

Quadriceps, 128, 129

Radius, styloid process of, 4
Rectus abdominis, 68, 84, 92, 93
border of, lateral distortion of upper part of, 92, 93
inscriptio tendinea of, 82, 83, 94
Rectus femoris, 116, 128, 144
Resistance of lever system, 1
Ribcage, 80, 81
abdominal wall and, 96, 97
Ribs, 70, 71, 80, 81, 82, 83, 94, 95
abdomen and, 72, 73
mobility of, 72, 73

Sacrospinalis, 44, 48, 52, 132
Sacrum, 42, 43
Sartorius, 84, 110, 112, 116, 118, 122, 144
Scapula, 44, 50
inferior angle of, 34, 36, 56
medial border of, 4, 36
spine of, 16, 38
Scrotum, 110, 116
Semimembranosus, 132, 140
Semitendinosus, 66, 114, 128, 132
Semitendinosus-membranosis, tendons of, 146, 147
Serratus anterior, 22, 58, 59, 66, 68, 82, 83, 170
obliquus externus and, digitation of, 84, 94
Shoulder, 26, 27, 28, 29, 34, 35
back and, 36, 37, 38, 39
chest and, 40, 41
arm, neck and, 170, 171
head and, 30, 31, 32, 33
head and, 20, 21
neck and, 16, 17, 22, 23
chest and, 24, 25
thighs, trunk and, 47, 49
Shoulder joint, 28, 29
adduction in, 26, 27
static horizontal, flexion-adduction in, 32, 33
Skeleton
bony landmarks of, 4
as structure of levers, 1
Soleus, 52, 128, 138, 140, 144, 158
Spina iliaca anterior inferior, 118
Spina iliaca anterior superior, 44, 66, 84, 118, 120, 121, 122
Spina iliaca posterior superior, 136, 138
Spinal column, spinous processes of, 42, 43
Spine
curvature of, in flexion, 58, 59
extension in, 64, 65, 66, 67
of ilium, posterior superior, 48, 49
of scapula, 16
1st thoracic, 38
7th cervical, 4, 38, 42, 44
Spinous processes of spinal column, 42, 43
Splenius capitis, ridges of, 38
Sternal attachments, 14
Sternoclavicular joint, 16
Sternohyoid, 14
Sternomastoid, 14, 16, 20, 24, 170
clavicular part, 22

Sternomastoid *(continued)*
sternal part, 22
Sternum, 24, 94
Styloid process
of radius, 4
of ulna, 4, 172, 180, 186, 187
Supraclavicular fossa, greater, 16, 17
Supraspinatus, 22, 52
trapezius over, 36
Suprasternal fossa, 14

Teen-ager (full figure of)
female
age 13 years, 217
age 15 years, 219
age 16 years, 221
age 19 years, 225
male
age 16 years, 223
age 19 years, 227
Tendo calcaneus, 138
Tendon(s)
of biceps femoris, 58, 59, 116, 128, 146, 147
of extensor digitorum longus, 144, 160, 161
of semitendinosus-membranosis, 146, 147
patellar, 128
Tensor fasciae latae, 110, 112, 116, 118, 122, 128, 130, 131
Teres major, 34, 36, 56, 68, 174
Thigh
abdomen and, venous pattern of, 108, 109
legs and, 128, 129
trunk, shoulders and, 47, 49
Thoracic spine, 38
Thoracic vertebrae, 42, 43
Thyroid cartilage, 16
Tibialis anterior, 144, 158, 162, 164, 166
Trapezius, 14, 16, 22, 36, 40, 44, 56, 170
upper border of, 38
Triceps, 22, 26, 36, 174
action of, on forearm, schematic representation of, 2
lateral head of, 176
long heads of, 170, 174, 176
medial heads of, 170
Trochanter major, 4, 118, 122, 126, 127, 136
Trunk, 44, 45, 60, 61, 62, 63, 82, 83, 84, 85, 86, 87
extension of, 64, 65, 66, 67
legs and, 118, 119, 120, 121, 122
neck and, 42, 43
of athletic male, muscle definition in, 82, 83
of nonathletic male, muscle definition in, 86, 87
shoulders, thighs and, 47, 49
Tuberositas metatarsalis V, 158, 160, 162
Tuberositas tibiae, 4, 116, 122

Ulna
head of, 188
styloid process of, 4, 172, 180, 186
Umbilicus, 112, 114

Vastus lateralis, 116, 128, 138, 140, 144
contour of, 136, 137
Vastus medialis, 116, 122, 128, 144
Vena basilica, 182, 183
Vena cephalica, 186
Vertebra(e)
lumbar, 42, 43
thoracic, 42, 43
Virginal breasts, 73, 74

Zygomatic arch, 14, 15